CAST METALS TECHNOLOGY

CAST METALS TECHNOLOGY

J. GERIN SYLVIA
Department of Industrial Engineering

The Pennsylvania State University
and American Foundrymen's Society Training and Research Institute

ADDISON-WESLEY PUBLISHING COMPANY
Reading, Massachusetts
Menlo Park, California · London · Amsterdam · Don Mills, Ontario · Sydney

ISBN 0-201-07395-1
CDEFGHIJK-MA-79876

To Dorothy T. Sylvia
My wife whose patience and cooperation
were necessary aids to my completion of this book.

PREFACE

In the last decade new levels of excellence in the cast metals industry have surpassed many of the skills of the artisan of yesteryear. The mechanization of a technical industry has been due, in part, to the application of the laws of physics, chemistry, ceramics, and metallurgy to the many foundry problems.

The results have been revealed in mass production of intricate shapes, thinner walled castings, dimensional stability of cast metals, new mold materials, improved melting equipment, high density molding, closer metallurgical control of metals, completely automated foundries and guaranteed castings. The technical occupations within the metal casting industry form a vitally important part of science and engineering, and they demand personnel with a background and education within the technical education system.

The purpose of this text is to provide a functional perspective for the technician whose job requires correct decisions, related to the application and limitations of casting processes. It is a comprehensive treatment of equipment, materials, and operations intended to provide a realistic understanding, rather than the development of a special craft skill, within the industry. It treats the subject matter at the technician rather than at the engineering level.

While methodology of numerous procedures and processes is discussed, it is expected that the student will apply scientific knowledge in his practical training during which he obtains a "feel" for his understanding and thus can reason the "why" for the various phenomena. The student should be trained to be conscious of the operating variables at play and to make observations which will help him to decide on the proper operating conditions.

The book is intended to appeal to persons in the educational field, at the sophomore year in college, technical institute, community college, and junior college levels. With selected sections of Chapters 6, 7, 8, and 9, the subject matter could be used for the proper teaching to high school level students of the principles of the casting processes. Many concepts are related to high school mathematics, chemistry, or physics.

Though a student practices many of the concepts set forth in this book, it is advisable that the student familiarize himself, through visits to foundries, with practical operations. By visiting foundries of different sizes with varying degrees of automated and mechanized operations, he would have a broad in depth understanding of the many aspects of cast metals technology.

Lakeville, Massachusetts   J.G.S.
March 1972

CONTENTS

ix

CHAPTER 1

INTRODUCTION TO THE CASTING OF METALS

HISTORY OF THE CASTING PROCESS

What is a metal casting? A metal casting may be defined as a metal object produced by pouring molten metal into a mold containing a cavity which has the desired shape of the casting, and allowing the molten metal to solidify in the cavity. If sand molds are used, the molds are destroyed upon solidification of the metal. If a permanent-type mold is used, it is merely separated to remove the casting.

Historical data indicate that metal casting had its beginning some four thousand years before the Christian era. There seems little question that metal was first used in the part of the world known as the Eurasiatic steppe belt (the Russian Black Sea area) (Fig. 1–1).

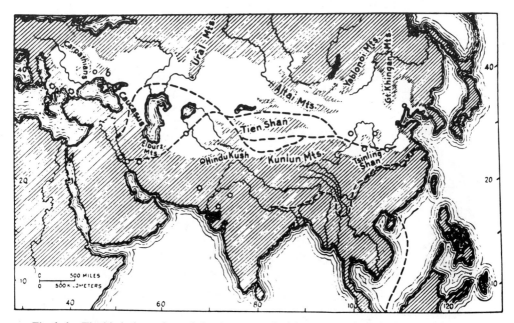

Fig. 1–1. The birthplace of metals has been traced with some accuracy to the steppe corridor of Eurasia, to the area north of the Black Sea in the Carpathian Mountains of Russia. (From B. L. Simpson [1].)

It is only natural that gold was the first metal to gain attention. It must surely have been evident in streams and waterways. Early man undoubtedly picked up pure nuggets and admired their excellent surfaces. It was soon discovered that gold was malleable, and could be flattened without splitting. At about the same time, early man discovered copper beads that had come from copper-bearing ore which had been used to bank fires. Copper could be shaped by heating and hammering, and thus—because it suited the purpose of early man—copper became the material he used to produce castable molten articles.

When the people of the Black Sea area swept down into Mesopotamia some four thousand years before Christ, conquering as they went, their victories were due primarily to their forged weapons. It was in their new habitat that some early foundrymen invented a high forge fire and produced a cast object from molten metal. Thus, before the wondering eyes of some early foundryman, the art of casting metal was born.

Copper was the first metal to be cast. Later it was noticed that sometimes copper contained certain other substances which made it harder; thus bronze was discovered. The material which had been accidentally combined with copper was tin.

Gradually artisans began to notice that the amount of metal produced was somewhat proportional to the air draft, which gave a hotter fire. Then an upright shell, lined with clay, was charged with alternate layers of copper ore and wood. Hollow wood blowers (tuyeres) were introduced into the furnace at the level of the firebox. These blowers were crude bellows covered with goatskin at the blower

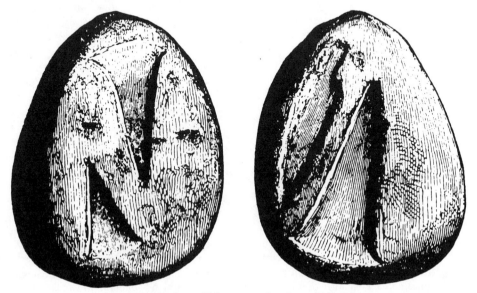

Fig. 1–2. Early mold for a spearhead, cut in fire clay.

end. With the aid of spring poles tied to each bellows, the operator stepped on each bellows, forcing air into the furnace. When he removed his foot, the bellows snapped back into normal position and air flowed back into them. This provided a fairly steady stream of air to the furnace, thus making it a blast furnace.

Many of the earliest molds were open half-molds, made of sand (Fig. 1–2). Later on the molds were cut in stone or fire clay. Simple shapes were formed to

Fig. 1–3. A ceremonial bronze elephant, cast in two parts and joined together; this is an example of the workmanship of early Chinese foundrymen. Probably from the Chou dynasty (1122 to 255 B.C.) or earlier. (From B. L. Simpson [1].)

produce spearheads, axheads, and simple agricultural implements. Melting was first done in clay-lined holes in the ground, but gradually, as time went by, artisans developed more permanent melting media, which eventually evolved into melting furnaces.

The early masses of people were migratory. Hence the casting process moved eastward into the Orient. It continued to be developed there until about the tenth century. The intricate and delicate work of the early Chinese foundrymen indicates the use of lost-wax techniques, closed stone molds, and sectional loam molds. The Chinese achieved a mastery of bronze casting and an advanced knowledge of metallurgy unknown in earlier times (Fig. 1–3).

Iron had been discovered as far back as 2000 B.C. Early man first shaped it by reducing the ore, melting it, and puddling it with slag into a ball, then forging it as wrought iron. The Chinese are known to have made castings of iron about 600 B.C. In India, cast crucible steel was first produced about 500 A.D., but the process disappeared and was lost until rediscovered later by Benjamin Huntsman, in England, about 1750.

After spreading eastward, the casting process spread westward into the Near East, the Mediterranean basin, and the rest of Europe. The Egyptians improved the already well-developed casting techniques of the Orient. They are also credited with the discovery of the lost-wax process of casting metal shapes. The use of a cope and drag, and of core molding, appeared early in the Egyptian artisans' development of metal casting (Fig. 1–4).

Iron was not generally cast in Europe until about the fourteenth century. Previous to this it was shaped by forging. The high temperature needed for pouring iron castings was obtained in furnaces resembling small blast furnaces. The air necessary for developing the melting temperatures was supplied by large bellows operated by hand, foot, or water power. Iron ore was reduced in these furnaces, and, when it was molten, it was allowed to flow directly into molds.

At the beginning of the thirteenth century the chief interest of the foundrymen was the casting of bells for the large cathedrals then being built in Europe. The molds for these bells were very often made in the churchyard or somewhere else near the cathedral. The furnace for melting the metal was constructed alongside the mold. The bells were made by a process known as *loam molding* or *sweep molding* (Fig. 1–5).

According to history, the first cannon was cast in bronze in the year 1313 by a monk in the city of Ghent. Artistic articles and statues were made in Italy by Benvenuto Cellini, who used loam molding combined with the lost-wax process. Leonardo da Vinci is also credited with the casting in metal of many fine works of art.

Vannoccio Biringuccio (1480–1539), the first true foundryman, can be called the "father of the foundry industry." In 1538 he became head of the Papal foundry in Rome. He is the first man known to have set down foundry practice in writing, in great detail. His clarity of analysis, together with a common-sense practical

Fig. 1–4. Cast bronze cat (made with core that had been removed). Cast in Egypt, probably in Sakkarah, in the seventh century B.C.

Fig. 1–5. Sweeping cope of bell mold in pit. Note complete core and cope, at left. Vertical sweeping, coupled with pit molding, was usually employed in molding bells. (From B. L. Simpson [1].)

approach, mark him as an accomplished artisan. Biringuccio's *Pirotechnia* undoubtedly covers all that was known of metallurgy in the sixteenth century. Even today, his statements of the three most important principles of making castings can go unchallenged. According to him, these principles are: "making and arranging the molds well, smelting and liquefying the materials of the metals well, and making the composition of their associations according to the results you wish to have." The exacting art of bell-casting bears out a point made by Biringuccio—that the art of casting is a "necessary means to very many ends."

Following the Renaissance, trade and commerce gradually revived, and the rise of a "free" industry brought into being the craft guilds. The guilds exercised complete control over all skilled workers, and exerted a great influence on foundry operations. The guilds accomplished a great deal of good by establishing principles of good workmanship, quality, and honesty. But because of their monopolistic power and the fact that they often carried rules to extremes, the guilds finally deteriorated.

In 1730, in England, a man named Abraham Darby, of Coalbrookdale, initiated the use of coke as a fuel. Thenceforward iron could be produced at about two-thirds the earlier cost, and thus coke became one of the principal tools of the iron foundries.   In 1794 there appeared the first metal-clad cupola, similar in appearance to those of today, invented by John Wilkinson of England. To provide the air blast needed, he used for the first time the steam engine invented by James Watt in 1765. Naturally enough, after the invention of the steam engine, the need for iron castings greatly expanded.

The first American foundry was established in 1642 near Lynn, Massachusetts, on the Saugus River. It was known as the Saugus Iron Works. The first American casting, the famous "Saugus pot," is the treasured property of the city of Lynn (Fig. 1–6). The Saugus area bog ore proved suitable for the start of an industry which eventually was to include more than five thousand plants in the United States. In quick succession plants were established along the Eastern seacoast as far south as Virginia.

Fig. 1–6. The first American casting. The iron pot known as the "Saugus pot" was made at the Saugus Iron Works, Saugus, Mass.

No history of the American foundry industry would be complete without a reference to the iron plantations, great estates which existed principally in Eastern Pennsylvania in the eighteenth century. Mount Joy Forge, later known as Valley Forge (of Revolutionary War fame), was one of many that started in 1742.

Paul Revere, the Revolutionary patriot who made the ride from Boston to Lexington on the night of April 18, 1775, to warn of the approach of the British, was a foundryman by trade. He operated a bell-and-fittings foundry in Boston. His success in metallurgy is well known today, and a great American company bearing his name—Revere Copper and Brass Company—is the direct descendant of Revere's original enterprise.

An important development of the nineteenth century involved the introduction of chilled-iron car wheels. This was, of course, of great importance to the railroad industry. In 1847, Asa Whitney, of Philadelphia, obtained a patent on a process for annealing chilled-iron car wheels cast with chilled tread and flange. Cold-blast charcoal iron was first used. Later a small amount of ferromanganese introduced directly into the ladle produced a chilled-iron car wheel which performed excellently. This made possible long hauls and heavier railroad freight loads.

A description of iron, to be complete, needs a mention of American "blackheart" malleable iron, as contrasted with European "whiteheart" malleable iron. Seth Boyden of Newark, New Jersey, is credited with much of the progress and growth of the American malleable-iron industry.

In attempting to duplicate European "whiteheart" malleable iron, Boyden experimented with an iron containing a larger percentage of silicon than was available in the European product. He produced a strong iron which could be machined easily. In the process, he managed to shorten the time of anneal to between six and ten days.

Steel castings appear to have been made first in India, and may have been poured in England as early as 1609. Inadequate equipment, however, held back the manufacture of steel castings in large quantities.

The introduction of the converter, the open-hearth furnace, and finally the electric furnace made it possible to produce steel commercially in great quantity, and to do so economically. Development continued, and in 1831 William Vickers of England was able to make cast steel from wrought-iron scrap by combining manganese oxide and carbon. Cast-steel guns were made by the Krupp Works in Germany in 1847.

In the United States, cast steel was produced by the crucible process in 1818 at the historic Valley Forge foundry. It wasn't until 1831 that William Garrard of Cincinnati, Ohio, utilized the excellent refractory clays of Cumberland, West Virginia, and established the first commercial crucible-steel operation in this country.

In 1851 William Kelly, of Kentucky, invented a converter which enabled him to produce a rather soft steel. Sir Henry Bessemer, in England, developed the process which bears his name. He succeeded in purifying the metal and assuring the presence of enough carbon to make steel. The first Bessemer converters in the United States were installed in Troy, New York, in 1865.

With the development of the open-hearth furnace in 1845 and the perfection in 1857 of the regenerative open-hearth furnace, with its great heat, the steel industry was given the tonnage capacity required for successful operation. The first open-hearth furnace in the United States was installed in 1870. Thus a vital tool for the growth of a nation came into being.

THE METAL-CASTING INDUSTRY TODAY

Gray iron, malleable iron, steel, copper, bronze, brass, aluminum, and other metal alloys became the metals which foundrymen were able to cast to shape. To complete the development of the art and science of casting, artisans in many lands made constant improvements in mold materials, sand conditioning, melting and metallurgy of metals, and handling of materials. The words "technological revolution" might be used to adequately describe the tremendous mechanization, automation, and system of controls that have grown with the industry, to develop it into the sixth-largest manufacturing industry in the United States today (Fig. 1–7). The manufacturing processes are outlined in Fig. 1–8.

Metal Casting: A Shaping Process

The production of various shapes of metal objects is accomplished by the pouring of molten metal into a mold, utilizing the flowability of liquid metals.

Types of foundries

The various types of foundries, based on the type of metal cast, are as follows.

1) *Iron foundries*
   a) Gray iron, producing high-carbon ferrous alloy.
   b) White iron, producing medium-carbon ferrous alloy.
   c) Ductile iron, producing a spheroidal-graphite ferrous alloy.
   d) Alloy gray iron, producing a variety of irons containing special-purpose alloying elements.

2) *Malleable-iron foundries*
   a) Malleable iron, producing an annealed white iron with graphite in nodular or temper form.

3) *Steel foundries*
   a) Carbon steel, which is a relatively low-carbon ferrous alloy.
   b) Alloy steel, which is any steel with appreciable amounts of special alloying elements.

(a)

(b)

Fig. 1–7. (a) A mechanized pouring line. Man at console controls movement of molds and pouring of molten metal. In the foreground, molds are coming from an automatic molding machine, being swung around, and placed in a horizontal position in the pouring line. (b) Hydraulic controls of an automatic molding machine.

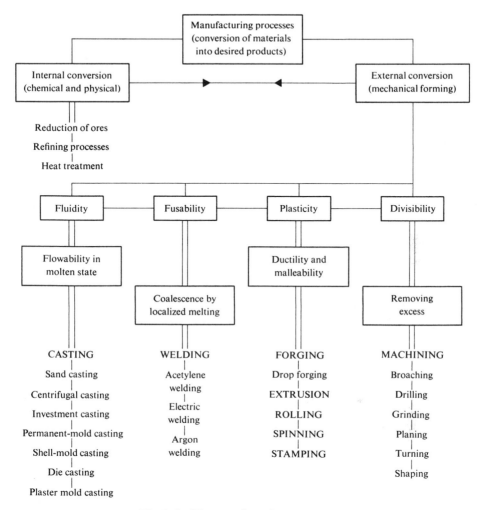

Fig. 1–8. The manufacturing processes.

4) *Nonferrous foundries*

   a) Brass and bronze, producing alloys with copper as the base metal, and other alloying elements.

   b) Aluminum, producing a variety of alloys, with aluminum as the base metal, and other alloying elements.

   c) Magnesium, producing a variety of alloys, with magnesium as the base metal (Dowmetal is an example).

Foundries may be classified in one of two categories.

1. *Jobbing foundries* (often independently owned) that contract for any job within their capability.

2. *Captive foundries*, which are usually departments of a larger manufacturing concern. Their castings are produced exclusively for the parent company. Many captive foundries, such as those found in automotive plants, achieve high production, and some sell a portion of their output.

Foundries may also be classified according to the molding process used.

a) *Sand casting.* A molding process in which a sand aggregate is used to make the mold; metal poured into the sand molds produces sand castings.

b) *Permanent-mold casting.* A process in which permanent molds made of steel or cast iron are used to receive the molten metal.

c) *Die-casting.* A process in which molten metal is poured under pressure into metal molds.

d) *Investment-casting.* A process sometimes known as *lost-wax* or *precision casting* which utilizes an expendable pattern of wax, plastic, or frozen mercury invested in a refractory material. When the pattern is melted out or volatized, molten metal is poured into the mold.

e) *Ceramic-mold casting.* A reusable pattern is used for this process, with a refractory slurry mold coat that is allowed to gel before the pattern is removed. After the mold is fired, molten metal is poured into the heated mold.

f) *Full-mold process.* A casting technique based on a foam plastic (a polystyrene type) pattern placed in a one-piece molding box. It is left to be gasified by the molten metal. The casting does not display any joint marks or flash and is accurately made.

g) *Plaster-mold casting.* A process of pouring molten metal into plaster molds or plaster-bonded molds.

h) *Centrifugal casting.* A process of pouring molten metal into a sand or metal mold revolving about either its horizontal or vertical axis.

Departments of foundries

Foundries are departmentalized, most of them having molding, core-making, melting, cleaning, and quality-control departments. Figure 1–9 is a flow chart of a typical operation.

The molding department is concerned primarily with the making of molds, and may use any one or a combination of the following: green sand, dry sand, bench, floor, machine, shell, pit, or loam molding.

Cores are sand shapes inserted into molds to form internal cavities. Core-making may be done by hand-ramming, machine-ramming, or air-blowing. Cores

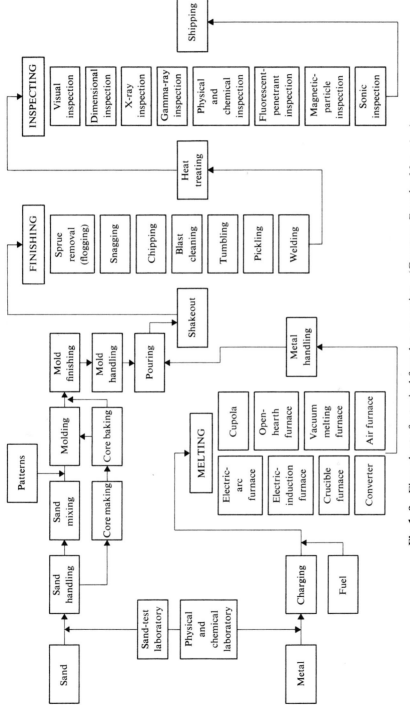

Fig. 1–9. Flow chart of a typical foundry operation. (Courtesy *Foundry Magazine.*)

may be made of oil-bonded sand, silicate-bonded sand, thermo-setting resin-coated sand, or by air-setting, or by the hot-box method. Some cores require baking and a storage area, while others are made and used within a short period of time.

The melting department is concerned with one or more of the following types of furnaces: cupola, open-hearth, air furnaces, electric furnaces, and crucible furnaces. Gas, oil, coke, pulverized coal, or electricity are fuels used in melting.

The cleaning department is concerned with removal of gates and risers, and the chipping, grinding, welding, shot-blasting, or tumbling of the castings, so that all sand clinging to the surface is removed. Surface condition is of course of paramount importance, and great care is taken to give the the casting a clean appearance.

The quality-control department is responsible for the control of metal analysis and castability, so that certain standards of physical properties are maintained. Control of the mold materials is necessary so that sand characteristics will produce sound castings. Control of the dimensions of all castings, as well as surface finishes which are sound and free of any voids or defects, is the goal in this department. To obtain guaranteed castings, nondestructive testing techniques are utilized, since closer tolerances and uniform quality are constantly being demanded by industry.

BIBLIOGRAPHY

1. Simpson, Bruce L., *Development of the Metal-Casting Industry*, American Foundrymen's Society, Des Plaines, Ill., 1948
2. Ekey, D. C., and W. P. Winter, *Introduction to Foundry Technology*, McGraw-Hill, New York, 1958
3. Aitchison, L., *A History of Metals*, Vol.I, Interscience Publishers, New York, 1960

CHAPTER 2

PATTERNS AND RELATED DESIGN

A *pattern* may be defined as a full-size model of shrink rule measurements used to produce a mold cavity into which molten metal is poured to produce a casting. Therefore, it is necessary to have a pattern whether a single casting or a great number of castings are to be made. It is important to have a suitable pattern, for the quality of the casting is influenced by the quality of the pattern. We shall not discuss the craft of patternmaking herein, but rather the fundamental principles that anyone who does any casting should know.

The number of the castings to be made from a pattern is the criterion for judging whether the pattern is to be made of wood (Fig. 2–1), metal, plaster, plastic, or perhaps styrofoam. Soft pine can be used when only a few castings are needed. For a longer pattern life and a sharper impression in the sand, a hard wood may be used. Metal patterns (Fig. 2–2), of course, are superior to either of the above materials. The end results, however, depend on the care of the designer and the precision of the patternmaker.

The patternmaker must have both skill and vision to interpret the blueprint and see the finished product. In order to communicate with the designer and the foundryman, he must understand the problems of casting. He must have a background and a working knowledge of mechanical drawing, and of plane and solid geometry, as well as a knowledge of the woods, metals, and finishes required in order to transmit an idea into a physical part.

Northern white pine, sugar pine, Idaho pine, and mahogany from Central America or Mexico are the materials most commonly used to make patterns. Aluminum, brass or bronze, gray iron, steel, and low-melting alloys are also often used, when one wishes to prevent excessive wear of the patterns, and when one needs the pattern to have certain other properties; for example, patterns used in shell molding must be heated. Thermosetting plastics produce glossy surfaces that are both hard and wear-resistant; these plastics also have excellent working qualities. For precision casting (investment), expendable patterns are made of wax, plastic, or frozen mercury.

Patterns are also often made of a combination of materials used for special purposes. To improve wear qualities and strength, metal inserts are often used, as well as resin-impregnated materials.

A pattern layout is generally made full size; materials and dimensions are transferred from the pattern layout to the work. To this is added the shrinkage

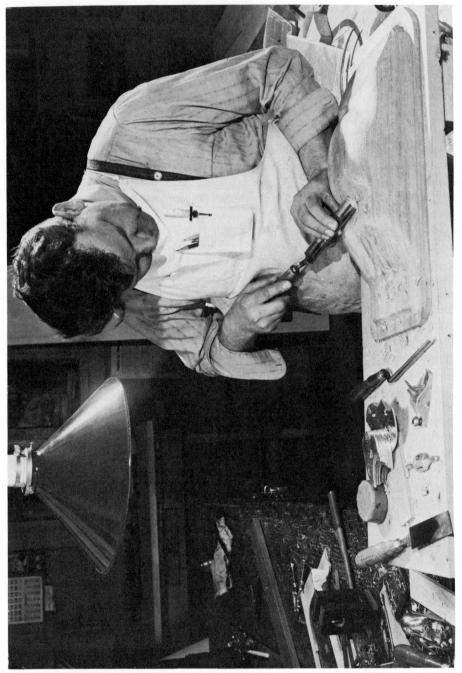

Fig. 2–1. Patternmaker working to close shape tolerances, carving the wood pattern of a special core plate pattern.

Fig. 2–2. A simple match plate of four rods, $1\frac{1}{4}$ in. in diameter by 14 in. long. Patterns are cast aluminum attached to a rolled aluminum plate.

allowance, as well as the draft or taper allowed on vertical faces of a pattern, so that it can be removed from the sand or other molding medium without tearing the mold cavity.

Shrinkage allowance on patterns is a correction for the metal's shrinkage during solidification and its contraction to its room-temperature size. The total contraction is volumetric, but the correction for it is usually expressed linearly. Thus *pattern shrinkage allowance* is the amount the pattern must be made larger than the casting to provide for total contraction. It may vary from $\frac{1}{8}$ inch to $\frac{1}{4}$ inch per foot, and it varies with casting design, type of metal, pouring temperature, and resistance of the mold to normal contraction of the casting. The linear allowances in Table 2–1 are representative of values for castings in sand molds. Special conditions prevail with some metals. White iron shrinks about $\frac{1}{4}$ inch per foot when cast, but during annealing it grows about $\frac{1}{8}$ inch per foot, resulting in a net shrinkage of $\frac{1}{8}$ inch per foot. Spheroidal graphite cast iron (nodular iron) may solidify with a contraction of $\frac{1}{4}$ to $\frac{1}{8}$ inch per foot, depending on the degree of graphitization which it undergoes during solidification.

The patternmaker uses a shrink rule to allow for the amount of shrinkage to be added to the pattern. For example, on a $\frac{1}{8}$ inch shrink rule, each foot is $\frac{1}{8}$ inch longer and each graduation is proportionately longer than the conventional length. Sometimes, if a pattern is first made in wood and then in some other metal (as one does in making master patterns), double allowances are made. An aluminum pattern made from a wood master pattern might require a total allowance of $\frac{1}{4}$ inch per foot on the wood pattern if a gray-iron casting is to be made. The total

Table 2–1 Pattern shrinkage allowances*

Casting alloys	Pattern dimension, in.	Type of construction	Section thickness, in.	Contraction, in. per ft.
Gray cast iron	Up to 24	Open		$\frac{1}{8}$
	From 25 to 48	Open		$\frac{1}{10}$
	Over 48	Open		$\frac{1}{12}$
	Up to 24	Cored		$\frac{1}{8}$
	From 25 to 36	Cored		$\frac{1}{10}$
	Over 36	Cored		$\frac{1}{12}$
Cast steel	Up to 24	Open		$\frac{1}{4}$
	From 25 to 72	Open		$\frac{3}{16}$
	Over 72	Open		$\frac{5}{32}$
	Up to 18	Cored		$\frac{1}{4}$
	From 19 to 48	Cored		$\frac{3}{16}$
	From 49 to 66	Cored		$\frac{5}{32}$
	Over 66	Cored		$\frac{1}{8}$
Malleable cast iron			$\frac{1}{16}$	$\frac{11}{64}$
			$\frac{1}{8}$	$\frac{5}{32}$
			$\frac{3}{16}$	$\frac{19}{128}$
			$\frac{1}{4}$	$\frac{9}{64}$
			$\frac{3}{8}$	$\frac{1}{8}$
			$\frac{1}{2}$	$\frac{7}{64}$
			$\frac{5}{8}$	$\frac{3}{32}$
			$\frac{3}{4}$	$\frac{5}{64}$
			$\frac{7}{8}$	$\frac{3}{64}$
			1	$\frac{1}{32}$
Aluminum	Up to 48	Open		$\frac{5}{32}$
	49 to 72	Open		$\frac{9}{64}$
	Over 72	Open		$\frac{1}{8}$
	Up to 24	Cored		$\frac{5}{32}$
	From 25 to 48	Cored		$\frac{9}{64}$ to $\frac{1}{8}$
	Over 48	Cored		$\frac{1}{8}$ to $\frac{1}{16}$
Magnesium	Up to 48	Open		$\frac{11}{16}$
	Over 48	Open		$\frac{5}{32}$
	Up to 24	Cored		$\frac{5}{32}$
	Over 24	Cored		$\frac{5}{32}$ to $\frac{1}{8}$
Brass				$\frac{3}{16}$
Bronze				$\frac{1}{8}$ to $\frac{1}{4}$

*From *Cast Metals Handbook*.

allowance on the original wood pattern would then provide for shrinkage of the aluminum pattern casting and of gray-iron castings made from the aluminum pattern.

When one is constructing patterns for castings in which various points on the casting's surface must be machined, one should provide enough excess metal for all machined surfaces. That allowance, commonly called *finish*, depends on the metal used, the shape of the part, the size of the part, the tendency to warp, and the machining method and setup. Whenever possible, surfaces to be machined should be cast in the drag side of the mold. This helps to provide a cleaner casting surface on the bottom, while dross, oxides, or other impurities float to the top. Any shrinkage would also be on the top surface and not on the surface to be machined. If finished surfaces must be cast in the cope side of the mold, an extra allowance should be made.

Table 2–2
Machine finish allowances and blueprint markings*

Pattern size, in.	Allowances, in.		
	Bore	Surface	Cope side
Cast iron			
Up to 6	$\frac{1}{8}$	$\frac{3}{32}$	$\frac{3}{16}$
6 to 12	$\frac{1}{8}$	$\frac{1}{8}$	$\frac{1}{4}$
12 to 20	$\frac{3}{16}$	$\frac{5}{32}$	$\frac{1}{4}$
20 to 36	$\frac{1}{4}$	$\frac{3}{16}$	$\frac{1}{4}$
36 to 60	$\frac{5}{16}$	$\frac{3}{16}$	$\frac{5}{16}$
Cast steel			
Up to 6	$\frac{1}{8}$	$\frac{1}{8}$	$\frac{1}{4}$
6 to 12	$\frac{1}{4}$	$\frac{3}{16}$	$\frac{1}{4}$
12 to 20	$\frac{1}{4}$	$\frac{1}{4}$	$\frac{5}{16}$
20 to 36	$\frac{9}{32}$	$\frac{1}{4}$	$\frac{3}{8}$
36 to 60	$\frac{5}{16}$	$\frac{1}{4}$	$\frac{1}{2}$
Nonferrous			
Up to 3	$\frac{1}{16}$	$\frac{1}{16}$	$\frac{1}{16}$
3 to 8	$\frac{3}{32}$	$\frac{1}{16}$	$\frac{3}{32}$
6 to 12	$\frac{3}{32}$	$\frac{1}{16}$	$\frac{1}{8}$
12 to 20	$\frac{1}{8}$	$\frac{3}{32}$	$\frac{1}{8}$
20 to 36	$\frac{1}{8}$	$\frac{1}{8}$	$\frac{5}{32}$
36 to 60	$\frac{5}{32}$	$\frac{1}{8}$	$\frac{3}{16}$
Admiralty metal			
Up to 24	$\frac{1}{4}$	$\frac{1}{4}$	$\frac{3}{8}$

*From *Cast Metals Handbook*.

Table 2–2 gives data which one may use as a guide to machine finish allowances, unless other specifications are given. Other casting processes permit different finish allowances to be used.

A taper of $\frac{1}{16}$ inch per foot is common for vertical walls on patterns drawn by hand. Machine-drawn patterns require about 1° of taper. Even vertical walls 6 to 9 inches deep may be drawn by machine if the pattern is very smooth and clean, and if the drawing equipment is properly aligned. However, in the case of pockets or cavities in the pattern, considerably more draft is necessary if one is to avoid tearing the mold during withdrawal of the pattern.

There are several types of patterns used in the production of castings. They are as follows.

1) Single or loose patterns.

2) Matchplate patterns.

3) Cope and drag patterns.

4) Special patterns or pattern devices.

1) *Loose patterns* are single copies of the casting to be made. However, they incorporate the allowances and core prints necessary to produce the casting. They are generally made of wood, but may be made of metal, plastic, plaster, wax, or any other suitable material. Hand-molding is practiced with loose patterns; thus the process is slow and costly. Gates and risers are usually cut by hand, but these can also be constructed as loose pieces and molded with the casting. Drawing the pattern from the sand is also done by hand after the pattern is rapped to loosen it from the sand; consequently the dimensions of a casting may vary.

2) *A matchplate pattern* is used when large quantities of small castings are desired. The cope and drag portions of the pattern are mounted on opposite sides of a wood or metal plate making the parting line. Matchplates are also cast with pattern and plate cast as one piece in sand or plaster molds. The runner and gate are mounted on the plate, and whenever possible the riser pattern is also included. Matchplate patterns are used mainly to increase productivity; thus they are generally used with some type of molding machine. The improved production rate, together with the increase of dimensional accuracy of the casting, justifies the increased cost.

Figure 2–2 shows a simple matchplate. Figure 2–3 shows an irregular matchplate with pattern. The center of the plate is contoured to fit the irregular parting line. Rollover machines may be used with this type of operation, thus increasing production. Flask equipment used for machine molding differs from equipment used for hand molding, in that the flask slips off at the completion of mold-making and slip jackets may be used to prevent movement of the mold during pouring operations.

(a)

(b)

Fig. 2–3. An irregular matchplate pattern. (a) Cope side. (b) Drag side.

3) When larger castings are desired, the patterns are mounted on two plates. The *cope* half of the pattern is mounted on a cope plate, and the *drag* half is mounted on a drag plate. The molders work on opposite sides of molding lines on two molding machines, one making drags and the other making copes. Perfect register must be the rule with this type of operation.

4) *Special devices* are used with loose patterns having an irregular parting line. A follow-board or pattern bed (match) is often used in this case. The pattern bed serves to support the loose pattern during the molding of the drag half of the mold, and also establishes the parting surface when the match is removed. Figure 2–4 shows a hard-sand bed (match) supporting a lever-arm pattern. The term *hard-sand bed* originates in the method of construction of the mold. A flask is first rammed with sand and rolled over. The pattern is then carefully embedded in (cut into) the sand, with care being taken to cut and smooth the parting-line sand. This is the pattern bed. A drag half flask is rammed over this bed. The mold is inverted, the pattern bed lifted off and set aside, and the cope half of the flask is rammed over the drag. Then normal procedure for loose-pattern molding is followed. The bed or match can be used a number of times. Plaster beds have a much longer expected life than sand beds before deterioration of the plaster interferes with efficiency.

Hollowed-out boards which perform the same function as pattern beds are known as *follow boards*. They support the pattern in the first molding operations and form the natural parting line of the castings. The molding sequence with a follow board is quite similar to the molding sequence used with a pattern bed. The drag half of the flask is rammed around the pattern and follow board. The mold is then inverted, the follow board removed, and the cope half is rammed. The mold is then finished in the normal molding procedure. Figure 2–5 illustrates a hand wheel made with the aid of a follow board.

It is desirable to have solid patterns that are designed to draw readily out of the sand toward the parting line. Draw backs, loose pieces, and cores are often used to form internal or external cavities or projections on patterns that perhaps would not "draw." The angle block in Fig. 2–6(a) is an example of this technique.

Sometimes a loose piece deep in the mold cavity is hard to withdraw. To make it easier to withdraw the pattern, one may use a loose boss on the end, dovetailed to facilitate removal. A simpler way to make the angle box is to use a core as in Fig. 2–6(b). This eliminates the loose piece; it also eliminates the possibility of disturbing the mold while one is withdrawing a pattern piece from a deep draw. The loose boss can be redesigned, so that the face of the boss extends to the parting line and has adequate draft (see Fig. 2–6(c)). This makes it easier to withdraw the pattern without a loose piece.

Green-sand cores are used to obtain internal configuration, or—as in the case of a sheave pattern—a floating or false cheek of green sand is used. However, this requires more manipulation of the mold; it also results in a mold that is prone to

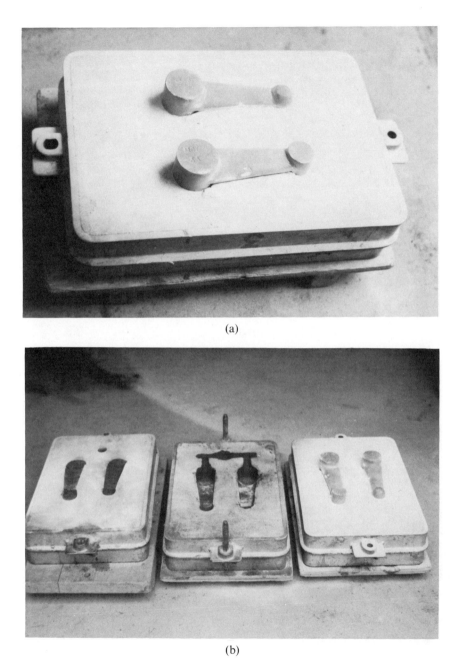

(a)

(b)

Fig. 2–4. (a) Green-sand match of two lever arm patterns used to make an irregular parting when ramming up a drag. (b) Green-sand match on right used to make drag (center) and cope (left).

(a)

(b)

Fig. 2–5. (a) Ten-inch hand-wheel pattern supported by follow board. (b) Pattern of a 10-inch hand wheel with a wooden follow board to fit a 14-inch-square flask.

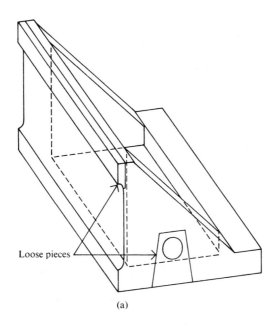

Loose pieces

(a)

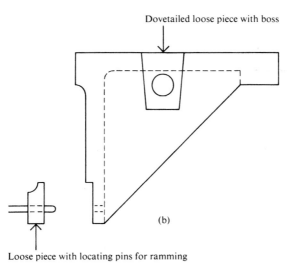

Dovetailed loose piece with boss

(b)

Loose piece with locating pins for ramming

Fig. 2–6. (a) The angle block; wood pattern with two loose pieces in place. (b) Angle block with loose piece necessitating deep draw plus dovetailed loose piece with boss. (See top of page 26 for continuation of Fig. 2–6.).

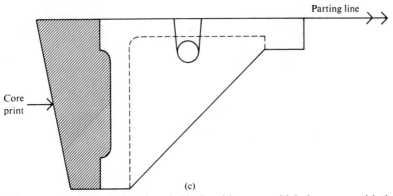

(c)

Fig. 2–6 (*continued*). (c) Redesigned angle block, with a core which does away with the need for a loose piece; the face of the boss is extended to the parting line to facilitate withdrawal of the pattern.

receive loose molding sand that cannot be easily removed. When this happens, the pattern should be redesigned to include a core print and a core box, so that a straight parting results, with all the mold in the drag. Using a ring core makes it relatively easy to make a sound, clean mold.

Figure 2–7 shows various steps in producing the mold, using a floating cheek or a ring core.

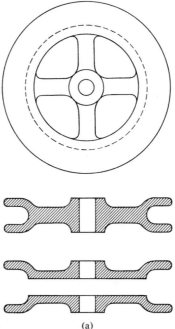

(a)

Fig. 2–7. (a) Sheave pattern. This is a split pattern that requires a floating green-sand core.

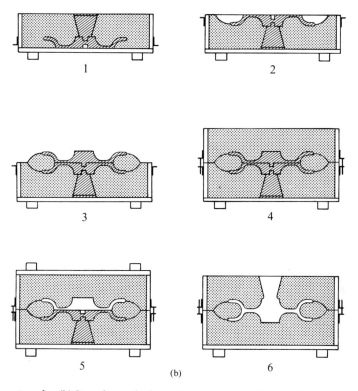

Fig. 2–7 (*continued*). (b) Steps in producing a sheave casting, using a split pattern and a green-sand floating core: (1) Cope section rammed up. (2) Cope rolled over, with irregular parting line cut. (3) Cheek sand rammed up between split pattern. (4) Drag rammed up against floating cheek and pattern. (5) Drag pattern removed and drag replaced. (6) Flask rolled over, cope half pattern removed, and cope flask returned. Mold has a floating green-sand cheek and is now ready for pouring of metal.

Plastic patterns may be made by injecting a plastic into a die to form a pattern. Or one can make a plastic pattern by machining a block of plastic. One can also utilize laminated construction by building up successive layers of resin and glass fiber. Another method is to pour plastic into a plaster mold either to make a new pattern or to duplicate an existing one, or perhaps to incorporate minor changes. Epoxy and phenol resins are the two plastics most commonly used for this work. Loose patterns or complete matchplates and patterns can be made entirely out of plastics.

Color markings for wood patterns have been recommended by the American Foundrymen's Society as follows.

1) Black—the body of the casting unfinished.

2) Red—surfaces to be machined.

3) Yellow—core prints and seats for loose core prints.

4) Red stripes on a yellow background—seats of and for loose pieces on the pattern.

5) Black stripes on a yellow background—stop-offs.

A *stop-off* is a portion of the pattern which produces a mold cavity that is later filled with sand, so that it doesn't become filled with metal during pouring and appear on the casting. A stop-off may be a reinforcing member on a frail pattern, the mold impression of the reinforcing member being filled in by a core.

Quite often one uses special sweeps instead of three-dimensional patterns to obtain shapes. Ordinarily these special sweeps are used when the shape to be molded may be formed by the rotation of a curved line element about an axis, as suggested by Fig. 2–8. The use of a simple sweep eliminates the necessity for constructing a large, expensive three-dimensional pattern. Occasionally a straight sweep is used to form a plane surface.

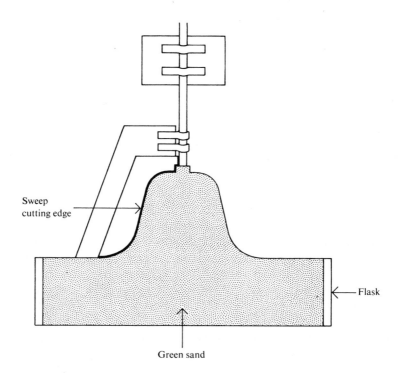

Fig. 2–8. Example of sweep molding. A green-sand mold of a bell is cut by a sweep cutting edge that is pivoted about a vertical axis.

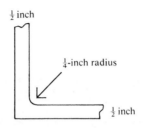

½ inch

¼-inch radius

½ inch

Fig. 2–9. Example of a fillet where two surfaces meet.

On all patterns in which two surfaces meet, one should use a small radius to join the surfaces instead of allowing them to intersect in a line. Such a radius is called a *fillet* (Fig. 2–9). Fillets are necessary in order to avoid shrinkage cracks at intersections and to eliminate concentrations of stress. They should be related in size to the thickness of the casting walls. Fillets are made of wood, leather, or wax, and may be shaped to the desired contour. They are either glued on, or, in the case of wax, pressed into the area with a heated fillet tool.

Laminated plastic patterns with surfaces as smooth as glass are more durable than wood patterns. They draw more easily, are not prone to wear and scratches in service, and can be repaired quickly when damaged. Plastic pattern materials also provide greater dimensional stability, better surface finish, and lower overall costs.

A laminated fiberglass core box can resist tremendous pressures and rough treatment in foundry practice. Various combinations of epoxys are employed, often with lay-on fiberglass for added strength. Metal grains are also added to the resins, to provide added strength to those areas likely to be abused by foundry methods.

The polyurethanes exhibit outstanding resistance to damage, both from abrasion and indentation.

Aluminum patterns and core boxes have the advantages of light weight, good machinability, good conductivity of heat, and generally satisfactory resistance to abrasion. Practically all the aluminum matchplates or cope and drag plates are cast in plaster molds. Smooth pattern surfaces and accurate reproduction of fine details in the pattern also make possible close dimensional tolerances when one is using plaster molds.

Aluminum alloy patterns and core boxes are frequently suitable for shell molding (Fig. 2–10). A good alloy composition for a pattern is aluminum plus 7% copper, 5% silicon, and 0.15% titanium. This alloy exhibits good machining characteristics without the need for heat treatment. When the pattern or core box is to be operated at an elevated temperature, however, heating the pattern casting to 400°F for four hours before machining helps reduce dimensional change.

Fig. 2–10. Patternmaker constructing an aluminum core box. Critical area of the core box is being machined with the use of a special cutter in a drill-press head.

BIBLIOGRAPHY

1. *Cast Metals Handbook*, fourth edition, American Foundrymen's Society, Des Plaines, Ill., 1957

2. Marek, C. T., *Fundamentals in the Design and Production of Castings*, John Wiley, New York, 1950

3. American Foundrymen's Society, *Patternmakers' Manual*, second edition, 1960

4. Rogers, H. S., "Laminated Plastic Pattern Equipment," American Foundrymen's Society Transactions, **75,** pages 292–293, 1967

5. Sicha, W. E., "Aluminum Pattern Castings," American Foundrymen's Society Transactions, **69,** pages 479–482, 1961

CHAPTER 3

MOLDING, COREMAKING PROCESSES, AND MATERIALS

Cast shapes are made in molds and reflect the type and condition of the molds into which the molten metal is poured. Good-quality castings cannot be produced from inferior-quality molds. The commercial success of a casting process may reflect speed of production, dimensional accuracy, surface finish, metallurgical considerations, or some other particular feature inherent to a process. Castings are often described or named by the materials or methods of production.

Silica sand is the principal basic molding material used by the foundryman, whether for iron, for steel, or for nonferrous castings. It is relatively inexpensive and is sufficiently refractory for steel foundry use. Castings produced in silica sand are commonly known as *sand castings*. The simplest process of molding is sand molding. For small castings, molding may be performed on a bench; for larger, heavier castings, molding may be done directly on the floor. This requires a skilled man trained in the art of molding, if a high quality and good quantity of castings is to be produced. But to produce molds at the rate of 60 and upwards per hour requires not a skilled molder but a machine operator. This operator may be a person who is untrained in the art of molding. Machine molding is the production method of utilizing the fundamentals of mold-making. In machine molding, the operations are performed mechanically.

One way of classifying casting is according to the materials used in making molds. The various types of molds for sand casting are as follows.

1. Sand casting
   a) Green-sand molds
   b) Dry-sand molds
   c) Loam molds (sweep molds)
   d) Pit molds
   e) Shell molds
   f) CO_2 molds
   g) Air-set or self-curing molds.

There is also a classification of processes which usually do not include sand.

2. Permanent mold-casting and semipermanent mold-casting.

3. Die casting.

4. Investment casting.

5. Centrifugal casting.

6. Plaster mold-casting.

7. Other special mold processes
   a) Full-mold process   b) Ceramic process.

All processes are certainly limited, but many have definite advantages over others, and their selection is based on specific factors of greatest importance (see Table 3–1).

Table 3–1 Comparison of casting methods

Factor	Type of casting process				
	Sand	Permanent mold	Low pressure	Die	Centrifugal (or centrifuge)
Metals processed	All	Nonferrous and cast iron	Nonferrous[a]	Nonferrous[a]	All
Commercial sizes:					
min.	Few oz.	$\frac{1}{2}$ lb	$\frac{1}{2}$ lb	Minute	
max.	Largest	300 lb	100 lb	100 lb	Over 25 tons
Commercial surface finish, μin.	300–600	150–1000	50–150	20–125	20–300[b]
Tolerance, first in./in.	$\frac{1}{16}$	$\frac{1}{64}$	0.010	0.006	0.010[b]
Porosity[c]	5	4	3	1 or 2	1 or 2
Minimum section thickness, in.	$\frac{1}{8}-\frac{3}{16}$	$\frac{1}{8}-\frac{3}{16}$	$\frac{1}{32}-\frac{1}{8}$	$\frac{1}{32}-\frac{1}{16}$	$\frac{1}{16}$
Tensile strength, 1000 psi[d]	19	23	25	30	25
Production rate, pieces/hr[e]	10–15	40–60	50–80	120–150	30–50
Mold or pattern cost, $[e]	300	2000	3000	5000	1500
Scrap loss[f]	5	4	3	2	1

[a] Iron and steel are die cast in refractory metal molds, but only in a small way.
[b] In metal molds.
[c] 1 is least porous and 5 most porous.
[d] For No. 43F aluminum alloy, as example. Reference: *Metals Handbook*, 1961 edition.
[e] Figures for production and tool cost are relative for a 3-lb aluminum casting of moderate complexity.
[f] 1 lowest to 5 highest.

SAND CASTING

Where castings are produced in large numbers, the molds are made on molding machines. However, in order to design castings, patterns, and all the component jigs, fixtures, and core boxes, one must understand the fundamentals of molding.

Making a sand mold involves the proper ramming of molding sand around a pattern. After the pattern is removed from the sand and the gating arrangement is completed, the mold cavity is filled with molten metal to form a casting. The sand used is a mixture of sand grains, clay, water, and other materials added for specific purposes or properties.

Green-sand molding begins with inverting a loose pattern on a mold board and placing a suitable-sized flask on the board, as illustrated in Fig. 3–1. Loose patterns are used when relatively few castings are needed. A large production rate would call for match-plate or cope-and-drag-plate patterns.

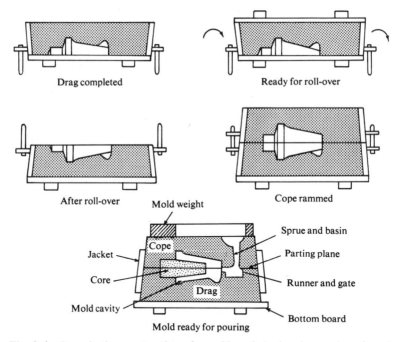

Fig. 3–1. Steps in the construction of a mold made by hand-ramming of sand.

The making of the mold requires the ramming of sand around the pattern within the confine of the flask. As the sand is rammed, hardness and strength are developed making the sand firm and rigid. This makes it possible for the walls of the mold to retain their shape and not collapse or erode when molten metal is poured into the mold cavity. Ramming of sand may be done by hand, as illustrated in Fig. 3–1. When ramming is done mechanically, it is performed by molding machines (we shall discuss these machines in Chapter 4). A *cope* (top half of the mold) and a *drag* (bottom half of the mold) may be molded in the same way, but the initial entrance of the metal through the sprue is provided for in the cope half of the flask. The gating and runner systems are usually with the mold cavity in the drag half of the flask. The gating system provides a means for molten metal to flow into the mold cavity. (Gating systems will be described in detail in Chapter 9.)

When cores are used in mold-making, they are usually set into the mold cavity of the drag half of the flask after the pattern has been withdrawn. *Cores* are sand shapes used to make internal and/or external configurations that cannot be obtained easily with the pattern alone. (The making of cores is illustrated in detail in Fig. 3–18.)

When the mold is finished, the cope section is placed on the drag section, thus closing the mold. The mold then needs to be clamped together, or a weight placed on the cope, so that the pressure of the molten metal will not lift the cope section and let the metal run out at the parting line.

GREEN-SAND MOLDING

Sand molds made with a moist sand, in which the moisture is present at the time metal is poured into the mold, are called *green-sand molds* (Fig. 3–2). Green sand

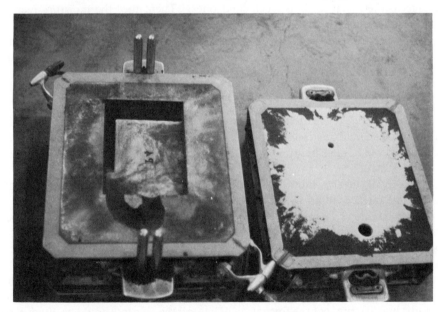

Fig. 3–2. A green-sand mold before molten metal is poured into it. Cope and drag are shown open. Mold is made in a slip flask. Light section of sand in cope is due to additive used in sand (see Chapter 4). Left: Drag half of the mold, with cavity and gating system visible. Right: Cope half of the mold, with sprue hole in foreground and vent hole near the back.

which may be used for making small, medium, and large castings, is used more extensively than any other mold material. It is relatively inexpensive to produce, since the basic material is readily available. Simple—even some complex—changes can be made in the pattern design, and thus in the mold design, at very little cost. Green sand is adapted easily and quickly to mechanical and automated production methods. However, green-sand molds are not as strong as some other type molds, and may be damaged with mishandling. Green sand requires moisture, and on account of this inherent characteristic, gas can be produced, and this can cause defects in the cavity and castings.

DRY-SAND MOLDS

A sand mold made with sand and a type of binder that does not require moisture to develop strength is called a *dry-sand mold*. This type of mold is commonly used for steel castings, but it may also be used for other types of alloys. Dry-sand molds are usually small to medium size. If such a mold is very large, it may be made in sections and then assembled. The binder used with the sand is a special oil which, when it is mulled, coats the grains of sand; when this sand is baked in an oven, the oil polymerizes, and the grains of sand are bonded together.

One advantage of dry-sand molds is that they tend to resist metal erosion; in addition, dry-sand molds are stronger and may be handled with less danger of damage to their shapes than green-sand ones. Then, too, the moisture-related problems of green-sand molds are not present.

Costs for dry-sand molds run higher than the costs of green-sand molds, however, due to the cost of the special oil binder and of the equipment needed to bake out the molds.

SKIN-DRIED MOLDS

Skin-drying is done to ensure a dry surface of the mold. Removal of water from the surface of the mold tends to eliminate pin-holing of a metal. The surfaces of green-sand molds may be dried with the use of a torch, hot air, or an infrared electrical lamp.

To control surface smoothness and to prevent the peeling of burned sand on the casting surface, mold sprays are used, which coat the mold surface with a refractory coating. These mold sprays are usually a mixture of water and a refractory material containing a bonding agent such as bentonite, cereal, or molasses. The coating may also be graphite which is dusted or brushed on the surface of the sand. Alcohol and other volatile liquids are widely used to replace about 90% of the water. When this is done, the surface of the mold may be ignited, producing enough heat to develop the necessary dry strength and to eliminate the need for torch-drying. To prevent the formation of gas, all the solvent should be allowed to burn off. However, since moisture in the backup sand will migrate through the dry skin, the molds must be filled shortly after they have dried. They cannot be stored for any length of time.

LOAM-MOLDING

Loam-molding is most suited for making large, heavy castings of a circular shape, such as large cylinders and paper-machine rolls cast on end. For this, a sub-structure is made of bricks, wood and other materials to produce the approximate contour needed for the casting. A slurry made of sand, clay, and water is daubed over the base structure, and then worked into a rough approximate shape. A sweep board is then rotated about the center axis of the mold cavity to finish sweeping the mold surface to the correct shape and size. The mold is then dried by torches or hot air.

No pattern is used in this case. The sweep board is so shaped that it produces the proper casting contours as it sweeps back and forth from a fixed guide or spindle. This eliminates the need for a pattern, and thus reduces the costs con-siderably. Loam-molding requires molders skilled in craftsmanship, however, since all the work must be done by hand.

Fig. 3–3. A pit mold that utilizes sections of cores to make a large pump housing. Workman places cores while standing in the center of the mold. Note positioning of the cores to form the mold.

PIT-MOLD MOLDING

Pit molds, which are used to produce castings too large for a flask, may be made in a pit by a bedding-in method (see Fig. 3–3). The pattern is set in a pit in the position in which the casting is to be poured, and sand is rammed or tucked under and around the sides of the pattern. The cope for the complete mold may rest on the drag at or above floor level, and may be bolted down to prevent run-out at the parting plane.

Many foundries have a concrete-lined pit equivalent to the size of the mold they customarily produce. The mold may be rammed up, striking off the surface to produce the desired shape. At times, when the design of the casting is such that a pattern cannot be drawn out of the mold, the entire mold cavity may be constructed with cores.

To prevent the forming of excessive internal stresses, large castings should cool slowly. Thus it may be days after these castings are poured before they can be subjected to air cooling.

SHELL MOLDING

During World War II a process was invented that came to be known as the *Croning* or *C-process*, after its German inventor, Johannes Croning. Now known as the *shell process*, it has several advantages over conventional molding methods. (1) Tolerances as close as 0.002 in. per in. have been held in some cases. (2) Less draft is required for shell molding than for sand molding. (3) Castings can be poured with thinner wall sections and with metal at a lower temperature than is possible in green sand. (4) Small-cored holes, often as small as $\frac{1}{4}$ in. in diameter and $\frac{1}{2}$ in. deep, have been accurately cast. (5) Much smoother surfaces can be obtained, due to the use of a very fine sand.

Some of the disadvantages of shell molding are: (1) The pattern equipment needed is more expensive than that needed for green-sand molding. (2) There is an added capital outlay for the shell-molding machine. (3) There is a size limitation. (4) The resin bond adds to the cost of the bonded sand, although in many cases this may be balanced off by the need for a shell as opposed to a flask completely filled with green sand. (5) A reasonably high volume of castings is needed to justify the selection of this process.

Fine sands are bonded with dry thermosetting phenol resins to produce permeable thin-shell molds. This bond is coated onto sand grains, which are dried at elevated temperatures. Patterns made of metal (usually iron) that are a part of a metal plate are sprayed with a liquid parting agent and heated to about 400 *!*F before being covered with the resin-bonded sand. The liquid parting agent keeps the pattern clean and prevents molding sand from adhering to the metal pattern. The sand is allowed to remain on the pattern for some seconds while the heat sets the resin and a shell forms over the pattern. Loose, uncured sand is dumped off. The shell is then cured in an oven for several minutes at about 600° F. The shell is now hard and is stripped from the pattern. The shell-molding process is illustrated in Fig. 3–4.

A complete mold can be made in two or more pieces and then pasted or clamped together. To add support or strength during pouring, the clamped shells are sometimes bedded in coarse sand, gravel, or metal shot. Two or more shell molds may be set up together in this manner.

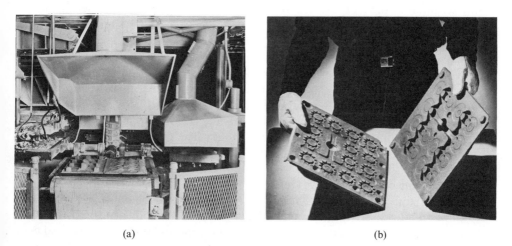

(a)   (b)

(c)   (d)

Fig. 3–4. Link-belt shell molding process. (a) Shell-molding machine with shell sand still on pattern plate. (b) Two halves which make one shell mold. (c) Half a shell mold with multiple castings from it. (d) Individual castings made from shell mold. (Courtesy Link-Belt Corp.)

Shell cores are made by using heated metal core boxes and curing the sand in the box. Delicately shaped cores are made in seconds; this leads to a tremendous saving in labor. (This will be illustrated in Fig. 5–9.) Specialized equipment is used for this process, which can be mechanized or completely automated to produce cores at very high rates of production.

CO_2 MOLDING

A description of the CO_2 *process*, by which molds and cores are hardened with carbon dioxide, was first published in various German and Eastern European journals in the early 1950's. Shortly thereafter foundrymen recognized the potentialities of the process and made rapid progress in developing it.

A typical technique for producing molds is to cover a pattern and fill the flask with silica sand mixed with sodium silicate. After the pattern has been rapped, CO_2 gas is blown through the porous sand mold for a short period of time. The CO_2 converts the sodium silicate to silica gel, and a hardened mold results. The pattern is then drawn from the hardened mold. The cope is made in a similar manner, and the mold is ready for immediate pouring of metal.

Cores are also made in this way, resulting in additional benefits from the process. Some of the advantages reported are: (1) No baking ovens or core driers are needed; thus one can obtain more capacity from the same floor space. (2) The castings obtained have more accurate dimensions and smoother finish. (3) CO_2 molding evolves less gas than many other core binders, and leads to a good shakeout of the casting. (4) Mold-drying is eliminated when CO_2 is used. (5) There is rapid production of finished cores, which expedites orders. (6) Economics: saving of space and time. (7) There is no odor or smoke from the process. (8) If one wants cores similar to shell cores, the cores can be hollowed out before gassing with CO_2.

This process can be adapted either to short-run or large-run production. It also facilitates the withdrawal of intricate shapes that are difficult to withdraw. Therefore this method is another important tool for the casting process.

CATALYZED CHEMICAL BONDING

Cores which are rammed and hardened by standing in air have core material composed of silica sand and a group of resins which when acted on by acids, form highly cross-linked polymers. The binders are made by mixing various proportions of urea, formaldehyde, and furfuryl alcohol. Furfuryl alcohol alone gives a relatively weak bond, and phosphoric acid increases the curing speed of the molds or cores.

Urea gives rise to pinhole porosity, and thus is not used in steel foundries for cold-curing or no-bake binders.

An acid used this way is referred to as a *catalyst*, and the most popular one is phosphoric acid. A typical sand mix is clean silica sand, 2% binder, and a catalyst which is between 30 and 40% by weight of the binder. The setting time of the cores is governed by the amount of catalyst and the temperature of the sand. The hotter the sand, the shorter the bench life of the mix. The amount of time it takes to strip the cores is important because the core boxes need to be released for another core, and the first core should be able to stand alone in air until it hardens completely.

Curing of the sand mix begins in the mixer as soon as the binder and catalyst come into contact. The bench life of the mix is short. When a continuous-mix slinger is used, the binder and catalyst are fed into the sand and the mixed core material is discharged directly into core boxes by using a slinger head. The discharge end is moved in an arc, so that a range of core boxes can be rammed in a short time.

In the hot-box process, sands are bonded with types of binders similar to those used in the cold-curing process. A weak catalyst is used, since the material is blown into core boxes which are heated to temperatures in the range of 400 to 500°F. Furfuryl alcohol polymerizes in the presence of weak acid under the action of heat from the core box. As in the case of the cold-curing binder, the alcohol and acid form a highly cross-linked polymer and the reaction is exothermic. This latter reaction is useful, since it promotes cold curing right through the center of the core. The procedure for mixing hot-box core material is similar to that for mixing cold-curing resin-bonded sands.

Since the mixed sand is used in core-blowing machines, it must have good flowability. The green strengths of the sand are low and the water content of the core material is critical. Too little moisture results in friable cores, while excess water increases baking time.

The efficiency of the process is decreased because the moisture of the sand must be eliminated or controlled; in addition, only a limited number of cores can be made in a given core box in a given time, since cores must cure within the boxes.

PERMANENT-MOLD CASTING

Castings are also often made by pouring molten metal into molds that are made of metal, or *permanent molds* (Fig. 3–5). Technically, a permanent mold is one into which molten metal may be poured more than one time without any variance in the dimensions or details of the castings. Generally speaking, the metal may be any molten metal; the mold material may be any material which meets the special refractory requirements; and the method of introducing the molten metal into the mold may be by gravity or pressure. In England, the permanent-mold casting process is called *gravity die-casting*. The die-casting process as practiced in the United States is referred to as pressure die-casting.

Permanent-mold casting is best suited to the production of castings in large quantities. As many as 200,000 low-temperature alloy castings have been produced in some molds before the molds have worn out. This type of casting may be classified as an intermediate between sand casting and die casting. Most non-ferrous alloys, as well as steel and certain gray-iron alloys, can be cast in permanent molds.

Fig. 3–5. Operator removing a gated casting from a permanent mold machine.

Gravity alone does not provide enough pressure to force molten metal into metal molds in sections much thinner than $\frac{1}{4}$ inch. Thus permanent-mold castings have cross sections much thicker than those of castings obtained by die casting. Permanent molds are usually made in two halves designed to open and close in alignment. The parting surface may be either vertical or horizontal. When the mold is closed, a cavity is formed, into which molten metal is poured. The metal is allowed to solidify and is then ejected. The mold material is usually graphite or a good grade of dense cast iron which can withstand quite high temperatures. The retention or radiation of heat which results from the continuous pouring of molten metal into the mold is important when one is planning how thick the wall of the mold is to be. The thickness may vary from $\frac{3}{4}$ to 2 inches, and usually follows the contour of the mold cavity.

Water cooling may be used if necessary. Since the mold is impermeable, venting must be accomplished by means of parting lines or ejection pins.

Cores for permanent molds can be made of metal, sand, or plaster. The casting design must be simple enough, and must have enough draft so that ejection from the mold is feasible. The life of the mold may be extended and the ejection of castings made easier by coating the cavity of the mold. To accomplish this, refractories suspended in liquid may be sprayed on the cavity. When the casting is to be made of iron, carbon soot from the flame of an acetylene torch is often used to coat the cavity of the mold.

Many castings require tolerances of ± 0.010 in., plus a good surface finish; this can be achieved by means of a permanent-mold casting. The chilling action of the mold also produces better metal properties in many alloys.

Molds made of graphite have been successfully used for nonferrous alloys, cast iron, steel, and titanium. The cavities of these molds are incised into graphite blocks by machine, in much the same way as is done for permanent molds. However, graphite soon shows signs of wear, and thus is used only in special cases. The thermal conductivity of graphite is about one-half that of carbon steel, and equal to that of stainless steel. Thus graphite molds have to be large, to absorb an adequate amount of heat. Wheels for railroad cars can be cast in graphite molds so accurately that no machining is required.

DIE CASTING

Die casting differs from permanent-mold casting in that the metal is forced into the mold cavity under high pressures, 1000 to 100,000 psi. This process is a highly mechanized one, in which dies may be interchanged without altering the machine. Half of the die is fastened to the stationary platen containing the gate. The movable platen containing the other half of the die is activated by a hydraulic piston which separates the two die halves at the parting plane. The casting is pushed out of the mold with ejectors installed in the movable die.

Two principal types of die-casting machines are used: *hot-chamber* and *cold-chamber* machines. The hot-chamber type is illustrated in Fig. 3–6. Molten metal flows into the chamber submerged in the melt and is then forced by a plunger into the die cavity. In the cold-chamber process the molten metal is hand-ladled into the shot chamber. The operation of the cold-chamber machine is illustrated in Fig. 3–7: (1) The metal is poured into the shot chamber. (2) The piston forces the metal into the die cavity. (3) The die opens. (4) The casting with the gate may be removed by automatic ejector pins. Nearly all castings have fins of flashing, where the metal has flowed between the die halves at the parting line. The flashing may be removed by press shearing (blanking) or by grinding. Lead, tin, zinc, aluminum, and magnesium alloys are commonly die cast. Some copper-base alloys are also die cast. Because of the high pouring temperatures of ferrous alloys, die casting of these alloys has become feasible only recently. It is achieved by using highly refractory molybdenum dies.

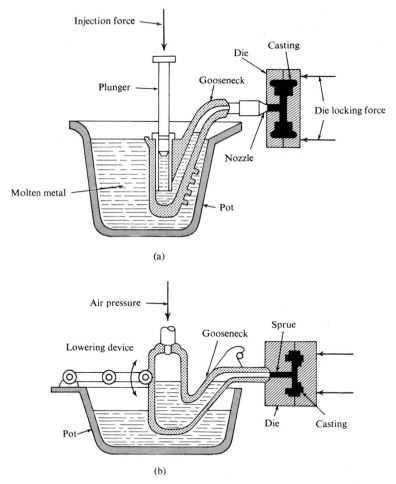

(a)

(b)

Fig. 3–6. (a) Hot-chamber die casting. (b) Air-injection die casting.

Metal for die casting may be melted in holding pots and transferred to the die-casting machine pot as needed. Dies are water-cooled to maintain them at constant operating temperatures. This prolongs the life of the die and provides the fastest allowable cooling rate for castings, so that they develop optimum properties.

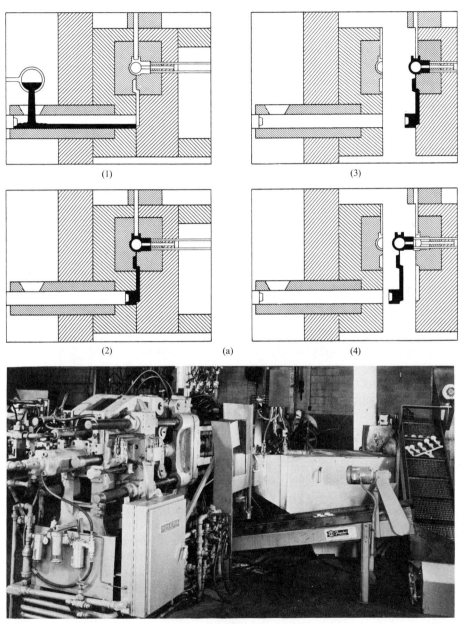

(1)   (3)

(2)   (a)   (4)

(b)

Fig. 3–7. (a) Cold-chamber die-casting machine. (1) Metal is ladled into chamber. (2) Plunger forces metal into die cavity. (3) Die opens. (4) Casting, together with gate and slug of excess metal, is ejected from die. (Courtesy Doehler-Jarvis Corp.) (b) Modern cold-chamber die-casting machine.

In comparison with other metal-working methods, die casting is done with a minimum expenditure of metal. Die castings are so accurate in size that very little or no subsequent machining is necessary after removal of the gate and flash. These castings are distinguished by their characteristic accuracy, smoothness, and good surface quality.

INVESTMENT CASTING

The *lost-wax* or *investment-casting process*, as it is known today, developed from an art that dates back as early as 1766–1122 B.C. during the early Egyptian and the later Shang Dynasty periods. In modern times, jewelers and dentists were the principal commercial users of the process until World War II began, when the investment-casting industry developed rapidly. This was due to the large demand for super-charger buckets required for aircraft reciprocating engines. Turbojet and jet engine parts are particularly amenable to this type of casting, and stimulated a rapidly growing industrial application of this process.

Investment casting uses expendable patterns of wax, plastic, or frozen mercury. Patterns are produced by injecting wax or plastic into a master die and removing them when they are solid. A number of patterns are then attached to suitable gates and risers, forming an assembly referred to as a "tree." (This is illustrated in Fig. 3–8.) The assembly is precoated by being dipped into a slurry of a refractory coating material. A suitable slurry consists of a silica flour (about 325 mesh) suspended in a water-ethyl silicate solution which has a viscosity that is sufficient to produce a uniform coating after drying. If precoating is not used, and if wax patterns are directly invested in the mold material, the mixture must be vacuumed to remove air bubbles that might lodge next to the patterns. Table 3–2 lists coating and investment formulas for investment molding.

The coated wax assembly is invested in the mold by inverting the assembly in a stainless-steel flask open at both ends. The investment-molding mixture is poured around the patterns, filling the flask. If the work table is vibrated, the mold material settles and completely surrounds the patterns. Table 3–3 lists typical investment-molding mixtures.

The molds are allowed to dry or air-set for six to eight hours. Then the wax is melted out of the hardened mold by heating it in an inverted position at 200° to 300°F. For burnout and preheating, the molds are heated at the rate of 100° to 160°F per hour up to 1200°F for aluminum alloys, 1600°F for brass, and to 1900°F for ferrous metals. The mold should be held at the temperature which is most desirable for pouring a given alloy and casting design. All wax and gas-forming material must be completely eliminated during the burnout and preheat cycle.

The mold may then be poured either statically or by means of centrifugal action. Almost any alloy that can be melted is amenable to investment casting if one adapts the refractory and mold temperatures to the requirements of the metal and casting design.

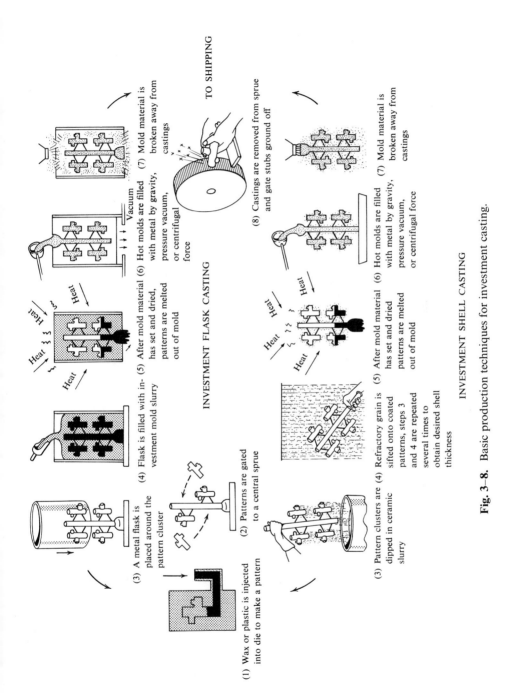

(1) Wax or plastic is injected into die to make a pattern

(2) Patterns are gated to a central sprue

(3) A metal flask is placed around the pattern cluster

(4) Flask is filled with investment mold slurry

(5) After mold material has set and dried, patterns are melted out of mold

(6) Hot molds are filled with metal by gravity, pressure vacuum, or centrifugal force

(7) Mold material is broken away from castings

INVESTMENT FLASK CASTING

TO SHIPPING

(3) Pattern clusters are dipped in ceramic slurry

(4) Refractory grain is sifted onto coated patterns, steps 3 and 4 are repeated several times to obtain desired shell thickness

(5) After mold material has set and dried patterns are melted out of mold

(6) Hot molds are filled with metal by gravity, pressure vacuum, or centrifugal force

(7) Mold material is broken away from castings

(8) Castings are removed from sprue and gate stubs ground off

INVESTMENT SHELL CASTING

Fig. 3–8. Basic production techniques for investment casting.

Table 3-2 Coating and investment formulas for investment molding*

No.	Material	Amount	Uses
1	Silica	67% Tetraethyl silicate, 8 parts by volume Water, 1 part by volume	Precoating for high-melting alloys
	Liquids	33% Ethyl alcohol, 1–2 parts by volume Hydrochloric acid, a few drops to 1 or 2% of 3% solution	
2	Solids	187 parts 94 parts 325-mesh silica 56 parts 325-mesh alumina 37 parts 40-mesh silica	Same as above
	Liquids	80 parts 4 parts 20 Be sodium silicate 1 part 2% polyvinyl alcohol	
3	Solids	60% plaster of paris 25% 50-mesh or finer silica 15% talc	Plaster molding for nonferrous alloys
	Liquids	Water to creamy consistency	
4	Solids	90% silica 6% magnesia 3% monobasic ammonium phosphate 1% monobasic sodium phosphate	Precoating for high-melting alloys
	Liquids	Water or 10% hydrochloric or nitric acid	
5	Solids	1464.4 g powder 3 parts china clay 17 parts 140-mesh silica flour	Same as above
	Liquids	800 ml 37.6% No. 40 ethyl silicate 59.8% 190-proof ethyl alcohol 2.6% hydrochloric acid in a 3% water solution	

*From *Principles of Metal Casting,* second edition, R. W. Heine, C. R. Loper, and P. C. Rosenthal, McGraw-Hill, 1967.

The basic method has many refinements. In a recently evolved technique, a thin ceramic shell is formed around the pattern. This shell is dipped into a rapidly setting refractory slurry, and then into a fluidized bed containing a grain-sized refractory. This is done six or eight times, with drying allowed between each dipping; this builds up a shell approximately $\frac{1}{4}$ inch thick. The baking and firing that follow produce a firm, accurate ceramic shell, ready to receive molten metal.

Table 3-3 Investment-molding mixtures

Refractory	Water	Binder	Uses
95% sand	27–31%	5% alumina cement	Investment molding
Sand		3% or more ethyl silicate or sodium silicate	Same as above, suitable for ethyl silicate precoat
91.2% sand	33.8%	6.5% primary calcium phosphate 2.30% MgO, 300-mesh	For ceramic or investment molding
90.6% sand	51%	7.1% primary calcium phosphate 2.3% MgO, 300-mesh	Same as above
93.3% sand	34.1%	5.17% primary ammonium phosphate	Same as above
1464.4 g		800 ml	Same as above
3 parts china clay 17 parts 140-mesh silica flour		37.6% No. 40 ethyl silicate 59.8% denatured ethyl alcohol, 190-proof 2.6% hydrochloric acid in a 3% water solution	

Figure 3-9 shows two 12-inch-high statues made by the ceramic-shell process. The principal advantages the investment-casting process has over conventional sand-casting and shell-molding castings are:

1) *Surface smoothness* (40–125 micro-inches, as compared to 80–165 micro-inches for shell-mold castings and 200–500 micro-inches for sand casting).

2) *Close tolerance* as cast. (This is why investment castings are often referred to as precision castings.)

3) *Minimum of machining* required for completely finishing castings.

4) *Capability of reproducing intricate shapes and sizes* that are difficult to cast or fabricate by other methods.

Some disadvantages of investment casting are:

1) The process does not lend itself well to mechanization; thus the volume of production by this method does not compare with that obtained with some other processes.

2) Investment casting requires an individual pattern for each casting to be made.

3) Investment casting requires numerous operations to complete castings, which makes the cost per casting greater than that of castings made by other processes.

The frozen-mercury process may be considered another kind of precision-casting, in that it affords a method of casting shapes that are more complex than those that can be cast by other methods.

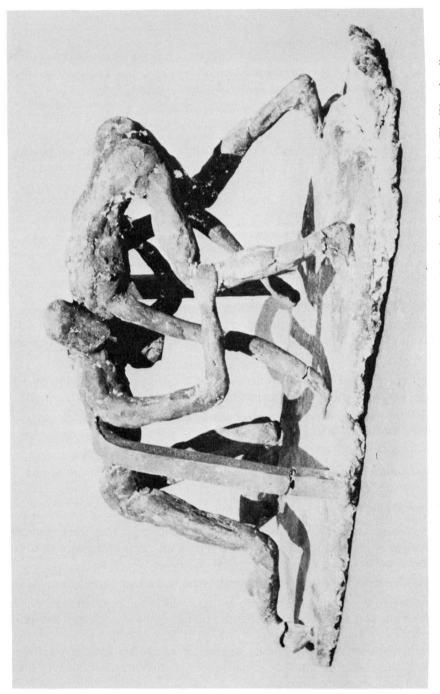

Fig. 3–9. Two ceramic-shell bronze castings made at the Wentworth Institute in Boston. (a) "The Wrestlers."

Fig. 3–9. (*continued*) (b) "The Gymnast."

CENTRIFUGAL CASTING

Centrifugal casting is accomplished by rotating a mold rapidly about its central axis as the metal is poured into the mold (Fig. 3–10). The centrifugal force distributes the metal within the mold and secures a denser metal with relatively few impurities such as oxides, slag, or gas. The process falls into three categories: (1) *true centrifugal casting,* (2) *semicentrifugal casting,* and (3) *centrifuging.*

Fig. 3–10. Brake drums cast by the centrifugal casting process.

1)   A pipe casting or a hollow casting with a central hole is probably the most familiar example of true centrifugal casting. The molds used for this process are usually metal molds or sand-lined tubular flasks. The DeLevaud pipe-casting machine in Fig. 3–11 illustrates the way a casting is made in a mold rotated horizontally. Cast-iron pipe is cast centrifugally in either metal or sand-lined molds, depending on the quantity desired. The casting machine is set up so that the mold can rotate at controlled variable speeds. A ladle is filled with the proper amount of metal for the casting of one pipe, and then the mold is moved into the pouring position. From the ladle the molten metal flows the length of a long pouring trough, which extends through the center of the mold to its opposite end for the start of the pouring. As metal starts to flow into the rotating mold, the mold is moved slowly away from the ladle. In this manner the metal is deposited on the metal mold surface in a helical path. After the pouring is completed, the rotation of the mold is continued until all liquid metal has solidified. A dry-sand core is often used to form the contour of the bell end of the pipe.

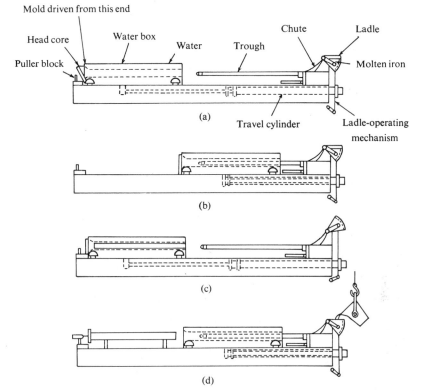

Fig. 3–11. Schematic diagram of DeLevaud pipe-casting machine and casting operations. (a) Head core put in place. (b) Start of cast. (c) Cast completed. (d) Pipe extracted, ladle refilled. (Courtesy H. W. Stewart.)

True centrifugal castings are made with either a horizontal, vertical, or inclined axis. Castings which are relatively short are more conveniently cast in a true centrifugal manner. The machine used for this process has to withstand strong forces, and thus should be strong and rigid. Horizontally rotated castings develop forces on the metal up to 75G (G = the force of gravity). Vertically rotated castings develop forces up to 100G. Thus, if slipping or raining of the liquid metal, as well as hot cracking, are to be prevented, rotating speeds must be carefully controlled.

2) Semicentrifugal casting is used for shapes that are symmetrical about a central axis; examples are wheels and gear blanks. It is not necessary that the castings have a central hole. A central sprue is usually provided and the castings may be stacked, as illustrated in Fig. 3–12.

The molds may be made of green sand, dry sand, metal, or some other suitable material. These molds must be rotated rapidly enough to develop a linear speed at the outer edge of the casting of about 600 feet per minute.

3)   Centrifuging differs from true centrifugal and semicentrifugal casting in that the entire mold cavity is spun off the axis of rotation. Metal is fed from a central sprue through a gate into the mold cavity. The molds are made with a number of relatively small castings arranged uniformly around a central sprue. Radial gates connect the castings to the sprue. Stacked molds may be used in this process, as illustrated in Fig. 3–12.

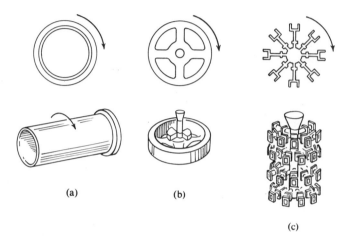

(a)   (b)

(c)

Fig. 3–12. Examples of (a) true centrifugal, (b) semicentrifugal, and (c) centrifuged castings.

PLASTER MOLDS

Plaster as a mold material for the production of nonferrous castings, most notably brass, was probably used 3000 to 4000 years ago by the Chinese. During the Middle Ages, the Italian craftsmen, Da Vinci and Cellini, used plaster as a mold material to produce their magnificent works of art. Since silver, gold, copper, and brass were the metals popular with the ancients, they were used exclusively to make statuary. The use of wax as a pattern material required investing in a hydraulic slurry rather than a rammed aggregate, as is used with core sand or green sand.

Plaster molds may be made in two or more parts, with a parting line, as in ordinary sand molding. Metal-casting plaster (100 parts of plaster added to 160 parts of water) is stirred slowly to a creamy consistency. It is important that the plaster be added to the water. If it is mixed very rapidly, too much air will be introduced to the slurry, and this will affect the mold texture. If it is mixed too slowly, the slurry will harden prematurely.

The slurry is then poured over a metal match-plate-type, pattern, or a loose pattern, and in a few minutes the plaster develops its initial set at room temperatures. The pattern can then be removed. The plaster is allowed to set. After setting, the molds are dried at 400°F. The mold sections are fragile, and require care in handling. Casting surfaces are extremely smooth, and have an accuracy of about 0.008 to 0.01 in. per in. Molten aluminum is usually forced into a plaster mold under air pressure to assure accurate reproduction.

Plaster mixes for metal castings contain 20 to 30% talc to prevent cracking of the mold, and sometimes contain terra alba or magnesium oxide to hasten the setting time. Occasionaly an additive is used to retard the setting time. Expansion of the plaster during baking can be controlled by adding small quantities of lime or cement. Plaster of paris, or calcined gypsum, is $CaSO_4 \cdot \frac{1}{2}H_2O$, and during the initial hardening reacts with the water of the slurry to form $CaSO_4 \cdot 2H_2O$. When it is dried at temperatures below 320°F, it reverts to $CaSO_4 \cdot \frac{1}{2}H_2O$. At temperatures above 320°F, the last combined water is driven off, leaving $CaSO_4$, or anhydrous calcium sulfate. All combined water must be removed, and care must be used to prevent the plaster from absorbing moisture after baking, because metal-casting plaster has a very low permeability.

The advantages of the plaster-casting process are that nonferrous castings which have thin sections and intricate shapes can be made with dimensional accuracy and excellent surface finish (Fig. 3–13).

A variation of plaster molding is the *Antioch process*, developed by Morris Bean at Antioch College in Yellow Springs, Ohio. By using investment methods, Bean made castings of frogs, fish, butterflies, and other creatures of nature. Low permeability of the molds was a constant source of trouble, however, and it required many experiments before this undesirable feature was eliminated. Finally, by means of autoclaving the mold in a moist atmosphere, Bean succeeded in casting a truly permeable mold.

The mold material used in the Antioch process is a mixture of sand, gypsum, asbestos, talc, sodium silicate, and water. The sand acts as a bulk material and the gypsum as a binder. In proportions of 50 parts water to 100 parts dry material, water is added to dry material consisting of 50% silica sand, 40% gypsum cement, 8% talc, and small amounts of sodium silicate, portland cement, and magnesium oxide. The pattern is set in a suitable flask and the slurry is poured over and around the pattern. In about 7 minutes, a set strength of about 70 psi in compression is developed. The mold is allowed to stand for 6 hours, and then closed and autoclaved in steam at about 2 atm pressure. It is then dried in the air for about 12 hours, and finally in an oven for 12 to 20 hours at 450°F. The autoclaving and dry process produces a permeability of about 25 to 50 AFS.*

Tolerances of $+0.005$ in. on small castings and $+0.015$ in. on large castings (such as tire molds) can be obtained. The Antioch process also makes possible better metallurgical quality in aluminium castings because metal chills can be embodied in the mold. An example of the intricacy and accuracy attainable by this process are the hydraulic torque-converter elements of automobile transmissions. The aerospace industry also uses the Antioch process to make many light-alloy castings because great accuracy and good surface finish are required. Figure 3–14 shows a radar wave guide made by the Antioch process.

---

* From American Foundrymen's Society.

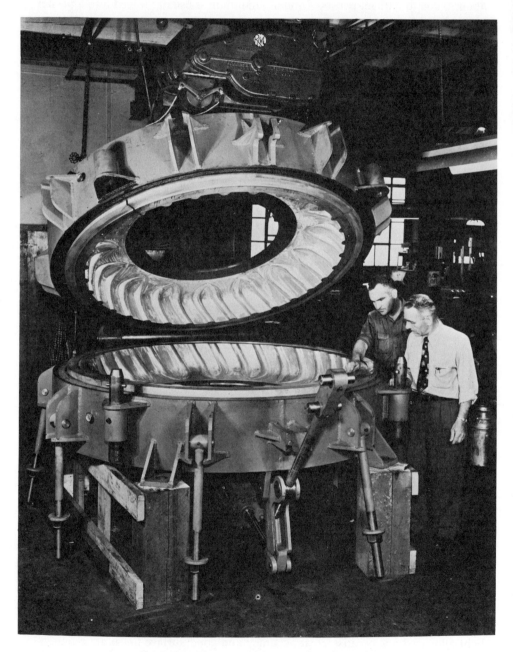

Fig. 3–13. Plaster mold for prototype tire mold.

Fig. 3–14. Aluminum casting made by the Antioch process: complex throat section for radar wave guide. Inside surface meets standing wave ratio requirements tested on several frequencies. Center wall terminal sections are 0.032 in. thick. Tolerance on individual openings is ±0.020 in., with same tolerance on combined openings. (Courtesy Morris Bean & Co., Yellow Springs, Ohio.)

SPECIAL MOLDING PROCESSES

In addition to the more common molding processes, there are several less frequently used methods and materials.

Low-melting alloys may be poured into molds made of silicone rubber or similar materials, to produce some types of castings. Many low-melting alloys of lead- and tin-base alloys are cast directly into such molds.

Rubber molds are quite simple to make and may be cast over a ceramic or metal pattern which retains the imprint of the pattern when the rubber hardens.

This method is economical for short runs and prototypes. Reproduction of fine detail is almost perfect. Complex and deep-undercut molds are easily produced with mold-making rubber. Objects such as toys, fishing lures, weights, and various jewelry and decorative pieces are easily cast.

FULL-MOLD CASTING PROCESS

The full-mold process is a molding and casting procedure utilizing gasifiable polystyrene patterns. With this process the pattern used consists either of one piece or of several pieces cemented together, and can be called a lost-pattern process, as such molds are made of foam plastics which vaporize as the metal enters the mold. Molten metal takes the place of the foam plastic pattern thus forming a cavity.

The principle of full-mold casting was disclosed in 1958 by an American patent published by H. F. Shroyer. To begin with, the technique was used for the development of art castings. It was not until 1962 that expanded polystyrene was utilized commercially to produce castings with a high degree of precision. Patterns including in-gates, runners, sprue, and riser systems are produced with foamed polystyrene, and are carefully molded in sand. Since the pattern is not to be removed, it is not necessary to make a two-part mold. It is advisable to face the pattern with a CO_2, air-set, or chemically bonded sand to avoid any possible deformation of the pattern during subsequent backup with green sand. Figure 3–15 illustrates a foamed polystyrene pattern fabricated from blocks of the material. The use of cores in the process can practically be eliminated, as the pattern need not be withdrawn. Thus any pockets, undercuts, or cavities may simply be filled with the cold-setting molding sand. The pattern may be formed in such a manner that it may serve as a core box at the same time.

This material may also be utilized as an addition to an existing wooden or metal pattern. The polystyrene is gasified and eliminated, thus leaving metal where metal is desired.

The patterns for the in-gate and feeder system can also be made in foam plastic. After the pattern has been embedded in the molding material, the mold is ready for casting. As the molten metal enters the mold, the pattern is progressively gasified. The molten metal takes the place of the foam-plastic pattern, and solidifies in this position.

Fig. 3–15. Complex polystyrene pattern made from multiple pieces of plastic material.

With the full-mold process, one can arrange ball feeder heads (blind risers) at every point of the casting one wants, without considering the mold parting line. The ball feeder head of polystyrene remains in the mold. Thus it can be located at any point at which feeding is most effective.

Many single castings are produced in considerable numbers by foundries utilizing this process. Practically all the automotive stamping dies used today are made by the full-mold process. Undoubtedly in the future the full-mold process will be used to mass-produce castings, for it has tremendous possibilities of improving existing castings while at the same time introducing new economies of production.

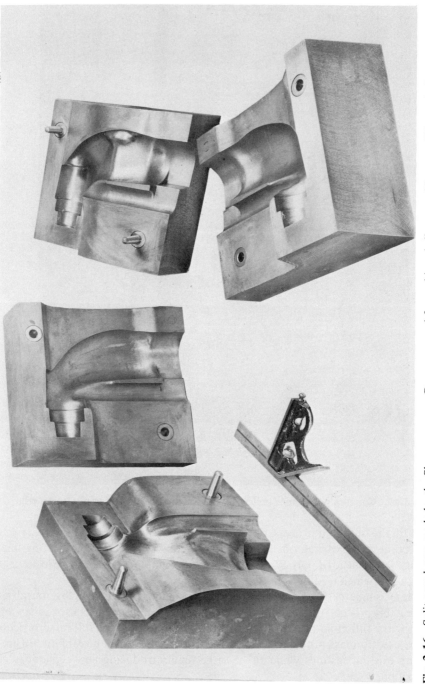

Fig. 3-16. Split core boxes made by the Shaw process.   Boxes are used for making shell cores.   (Courtesy Wisconsin Pattern Works.)

CERAMIC MOLDING; THE SHAW PROCESS

Technological changes have produced a potential investment-casting technique known as the *Shaw process* (after the English inventors Clifford and Noel Shaw). This process requires only standard foundry equipment, has no size limitations, and has wide applications, touching on the operation of almost every foundry and machine shop. Because of unique mold characteristics, the Shaw process is capable of producing castings with finer detail, smoother surfaces, finer grain structure, and greater accuracy and soundness than most other processes used today. There is less likelihood of porosity, gas, cracks, voids, or inclusions with this process than with many others.

A slurry-like mixture of refractory aggregate, hydrolized ethyl silicate, and a jelling agent are poured over a wood, plaster, or metal pattern. The mix is allowed to form a flexible jell over the pattern, and is then stripped off. The volatile portion of the mix is burned off with a torch and then the mold is brought to a red heat in a furnace. The result is a rigid refractory mold capable of receiving all metals, including steels.

On small castings, tolerances can be held within 0.002 in. per in., and up to 0.010 in. per in. on larger castings. Surface finishes can be up to 120 micro-inches. Split-core boxes made by this process are shown in Fig. 3–16.

The Shaw process originated in England, but is licensed to be used in the United States, where it has met with some variations. Ceramic slurry is used to produce solid ceramic molds that require four to five hours' baking time to remove trapped gases. A unique structure called *microcrazing* results from this heating. Microcrazing is a network of cracks which, under a microscope, looks somewhat like wire netting. It is three-dimensional, and each particle is separated by minute fissures or air gaps. The size of these fissures is critical, in that they must be small enough to prevent molten metal from entering them and large enough to accommodate the expansion of the ceramic particles when the mold is heated by the molten metal, and still provide adequate openings to vent air and gas. The air gaps between the refractory particles form an insulating mold structure with excellent heat capacity. This slows down the rate of feed demand and permits natural feeding. The result is sound castings with fewer risers.

CORE-MAKING

A *core* may be defined as a preformed mass of bonded sand that is used to make the internal configuration of a casting or to make the external contour that can be freed from the sand vertically as the pattern is withdrawn from the molding sand. In some cases the entire mold may be constructed of cores. Stacked molds may also be made in this manner (Fig. 3–17).

Cores are usually classified by the material from which they are made, i.e., green-sand cores, dry-sand cores, shell cores, CO_2 cores, furan-resin cores, and chemically activated cores. They are also classified by their position or use, such

Fig. 3–17. Stacked molds for producing crankshafts. (Courtesy Ford Motor Company.)

as horizontal, vertical, cover, balance, drop, ram-up, and others. Green-sand cores, which have relatively little strength, are made of green sand. Dry-sand cores are made with sand and binders which develop strength when baked. Besides being strong, cores must be collapsible, in order to shake out easily when the casting is dumped out of the mold.

The sand used as the base for the core may be selected from a number of sands such as silica, bank, olivine, zircon, and others. To meet the requirements of the castings produced, the sand should be selected for distribution of sand grains. fineness, shape, and base permeability. Thought should also be given to the fact that some of the core sand will be mixed with the molding sand during shakeout of the castings. The core sand should be readily assimilated without detrimental effects on the molding sand.

Organic binders such as cereal, resins, and core oils are the binders most commonly used. Cereal binders (corn flour, wheat flour, dextrine, starch) are used to provide core sand mixtures with both green and baked compression strength. These are commonly used with core oils (liquid organic binder). The resins that are used with the C-process and the hot-box process are also among the organic binders.

Among the list of inorganic binders are the three types of clay (Western bentonite, Southern bentonite, fireclay), sodium silicate (water glass) hardened by CO_2 gas, cement, chemicals, ethyl silicates, and various combinations of inorganic binders known by trade names. In general, inorganic binders in a core-sand mixture require much more water for tempering than organic binders.

Two major problems to be dealt with in the making of cores are gas and collapsibility. The source of gas in a core can usually be traced to an excessive amount of organic binders and/or moisture. These should both be kept at a minimum. Inorganic binders are potentially less gas-forming than organic binders, but they remain in the sand mixture at elevated temperatures and inhibit collapsibility. To a degree, this characteristic provides a core's resistance to erosion by the molten metal.

Cores are usually made in a core box which may be made of wood, metal, or other durable material (Fig. 3–18). The box may be a half box, split box, or booked type, to suit the case. Proper amounts of selected binder and sand are mixed in a muller so that the grains of sand can be uniformly coated with the oil. The core sand mixture is then rammed by hand into the core box (Fig. 3–19). After the surface is struck off flat, a metal or asbestos plate is clamped to the flat surface of the box. The box and plate are rolled over together, and the box is carefully lifted off, leaving the sand core on the plate. The sand core is then dried in an oven. Irregular-shaped cores may be supported by a contoured metal plate, called a drier, shaped to support the irregular core. The ramming of the sand may be by hand-ramming, jolting, squeezing, or blowing by means of a machine. Some cores may be made by extruding a core-sand mixture through a suitable die opening.

Perhaps the most common core-making machine is the *core blower*. Core boxes with vents to provide for easy escape of air are needed for the production of uniform cores. In this case the flowability of the sand requires careful consideration.

Fig. 3–18. Core box which has been given a hard chromium-plated facing on the mold surface. Workman has loose pieces of the core box that have also been chromium-plated to reduce wear.

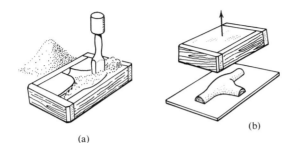

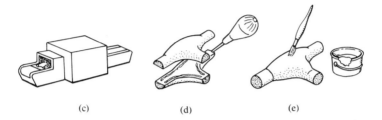

Fig. 3–19. Making a core. (a) Ramming core sand, (b) drawing the core box, (c) baking the core half (in a dielectric oven), (d) pasting the core halves, (e) "washing" the core with refractory slurry to improve finish of casting surface [2].

The cores are dried by being baked in some sort of oven. This equipment may be of simple design (shelves in a small gas-fired oven), or it may be many-shelved cars that roll into a large oil- or gas-fired oven, or it may be a belt which carries the cores through a dielectric or electric resistance unit. The baking time and temperature vary with the type of binder used, the amount of binder, the amount of moisture in the sand, and the size and shape of the core.

A core must be properly supported in the mold. Core prints on a pattern make recesses in the sand, thus providing for the proper positioning of the core. Sometimes *chaplets*—small metal props that fuse with and become a part of the casting (Fig. 3–20)—are used to support and position the core in the mold. They must be clean and dry, and should be made of the same type of metal that is to be poured into the mold. Some cores are made with rods or wires set in the core sand to provide additional support to the core in the mold.

Because the oil binder gives off a gas when it comes into contact with the molten metal, vent passages must be provided in the sand from the mold to the edge of the flask and into the atmosphere.

Figure 3–21 shows a large core that will require the use of chaplets to position it.

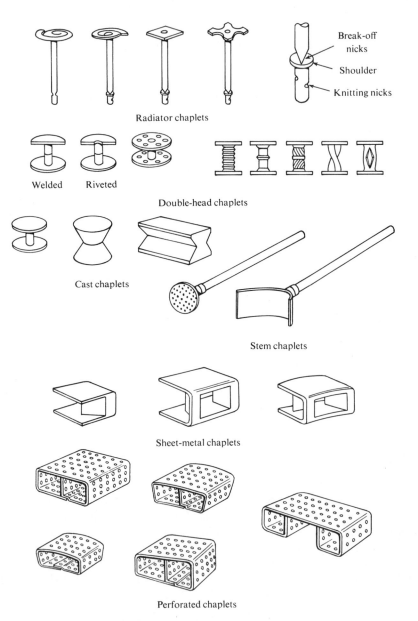

Radiator chaplets

Break-off nicks

Shoulder

Knitting nicks

Welded   Riveted

Double-head chaplets

Cast chaplets

Stem chaplets

Sheet-metal chaplets

Perforated chaplets

Fig. 3–20. Varied shapes of chaplets.

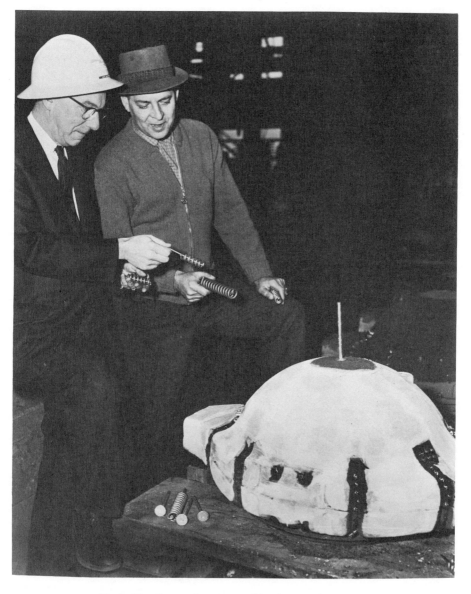

Fig. 3–21. Core to be supported by the use of chaplets.

BIBLIOGRAPHY

1. *Molding Methods and Materials,* American Foundrymen's Society, Des Plaines, Illinois, 1962
2. Sarkar, D. D., *Mould and Core Material for the Steel Foundry*, Pergamon Press, London, 1967
3. Ekey, D. C., and W. P. Winter, *Introduction to Foundry Technology*, McGraw-Hill, New York, 1958

CHAPTER 4

MOLDING SANDS

Sand can be defined as granular particles resulting from the disintegration or crushing of rocks or slag. Sand is an aggregate material essentially consisting of tiny, loose grains, minerals, or rocks which are no larger than $\frac{1}{12}$th of an inch or smaller than $\frac{1}{400}$th of an inch in diameter.

Sand also denotes a class of several minerals—rather than just one mineral— such as silica or quartz. Zircon, olivine, chromite, and ground ceramic minerals, as well as silica, are classified as sand when they are in this size range.

Silica sand is essentially SiO_2 (silicon dioxide) and is found in river deposits, lakes, and other large bodies of water. In many cases underground streams, no longer in existence, have left large deposits of silica sand. Silica sand usually comprises from 50 to 95 % of the total material in a molding sand.

Foundry sands of different types are found widely distributed over the United States. Sands such as zirconite and olivine offer special properties that often improve mold conditions. These sands are often more expensive than silica sand, and so are used for special applications. Such sands require the addition of a bonding material to make them molding sands.

There are two broad sub-classifications of foundry molding sand: *unbonded* and *bonded*. Washed and dried sand is unbonded mold or core sand, which may be composed of silica sand, olivine sand, or zircon sand as the base sand. A naturally bonded silica sand usually contains 70 % or more sand grains, with the balance being composed of clay, or the mineral bonding material may be thoroughly mixed with the sand grains. Naturally bonded sands are mined in many states, though the deposits of such sands usually are not very thick.

Molding sand, if it is to qualify as a good material, must be readily moldable and produce defect-free castings. The sand used must have several major characteristics that most other materials may lack. Chief among these are *permeability, cohesiveness,* and *refractoriness*.

Because steam and other gases are evolved in green-sand molding when molten metal is poured into the mold, the sand must be permeable—that is, porous—to permit the gases to pass off or else the casting will contain gas holes.

Another factor that contributes to the compactness or density of the sand grains is the shape of the grains. The grains may be rounded, angular, or subangular, depending on their geologic history. Typical shapes of sand grains are illustrated in Fig. 4–1. Molding sands used in foundries contain sand grains of

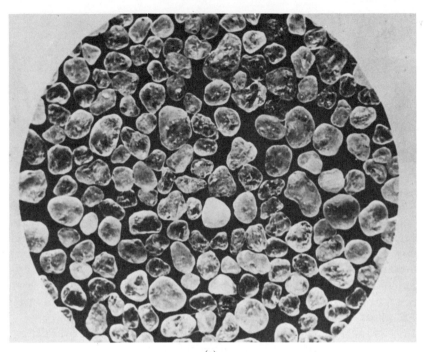

(a)

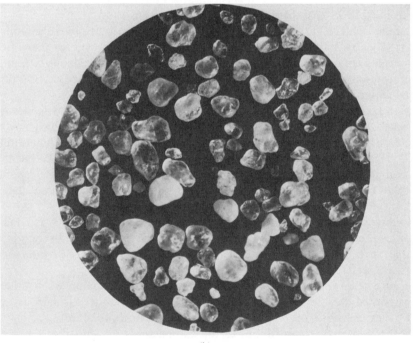

(b)

Fig. 4–1.　(a) AFS No. 40 grain subangular silica sand, 40× magnification.　(b) AFS No. 60 grain subangular silica sand, 60× magnification.　(Courtesy New Jersey Silica Sand Co.)

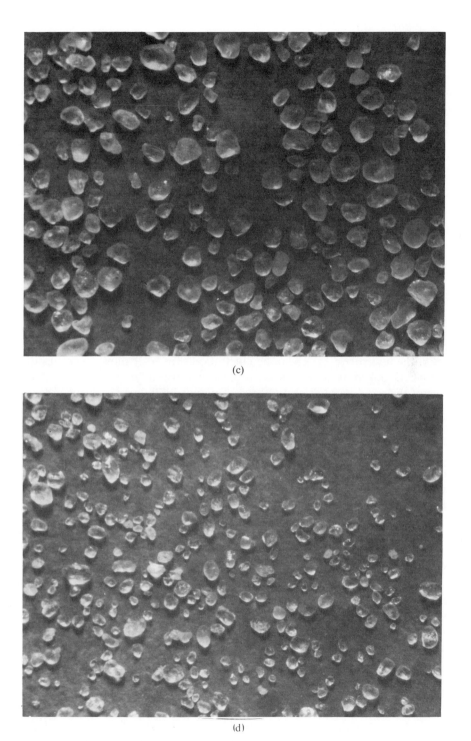

(c)

(d)

(c) Round-grain sand, <0.595 mm in diameter; 12× magnification. (d) Round-grain sand, <0.420 mm in diameter; 12× magnification. (Courtesy Ottawa Silica Co.)

mixed origin. Some are freshly introduced silica sand grains and others are grains of previously used molding sand, or parts of disintegrated cores.

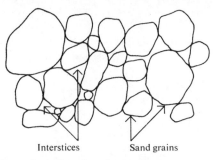

Interstices   Sand grains

Fig. 4–2. Schematic illustration of sand grains as they may appear under magnification.

The compactness of the sand grains affects the permeability of the mold on account of its effect on the interstitial structure. The moisture content in the molding sand also affects the permeability, because excess moisture tends to collect in the interstices. A schematic illustration (Fig. 4–2) shows how this may appear when looked at under a low-magnification magnifying glass. The bond content of the molding sand also affects the permeability in a similar manner.

The cohesiveness of the molding sand can be defined as the holding-together quality of sand grains. In order to control the properties of a molding sand, one must measure the strength of the molding sand in many ways. Green strength under compression, shear, transverse load, and tension are all measurable. Tests of green compression, in conjunction with tests of green deformation, are probably the most-used tests. Grain size and shape also affect strength characteristics, to the extent that a good molding sand must contain a proper mixture and distribution of grains of various sizes and shapes. The size and shape of sand grains has an important bearing on the packing of a mass of grains, on the interstices between the grains, and on the permeability as well as the flowability and deformation of the sand.

As one would expect, angular grains give greater mechanical strength than round grains, given the same amount of bonding material. Where maximum strength or bond is required, a subangular or angular type of grain—preferably with a roughened surface coated with a thin film of bonding material—is an advantage. Rounded grains give maximum flowability with high permeability. They ram easily, which is an advantage in mechanized core production. However, close packing of uniformly shaped grains encourages surface cracks and scabs on the face of the mold because silica grains expand when heated by molten metal.

The refractoriness of a sand is related to its ability to withstand high temperatures without breaking down or fusing. Impurities, such as metallic oxides, cause a lowering of the fusion point of some molding sands, and result in sand fused with metal at the casting surface.

To make core sand, one mixes a sharp silica sand (essentially free of binder) with selected binders, usually binders which have an oil base. Core-sand mixtures like molding-sand mixtures, must possess permeability, strength, and refractory qualities. In addition, cores must be collapsible and have smooth surfaces. Collapsibility of a core is related to the speed and degree to which the core breaks down after it has served its purpose in shaping the metal. This characteristic depends on the amount and type of binder, the temperature to which the core is heated when it comes in contact with the metal, and the length of time of contact. Surface smoothness is related to the degree of smoothness that a core imparts to a casting. The size of the sand grains and the proper use of core washes (coatings) are two big factors involved. A core sand with fine grains and with a thermosetting resin binder produces a casting with an extremely smooth surface.

It may be said that any material added to the sand—or provided by nature in the sand—which imparts cohesiveness to the sand is a *binder*. Clay is an example. Molding sands may contain from 2 to 50% of clay.

Table 4–1 Clay minerals used for bonding molding sands*

Clay mineral type	Composition type	Base exchange	Refractoriness (softening point)	Swelling due to water	Shrinkage due to loss of water	Particle size and shape
Montmorillonite Class IA, Western bentonite Source: Wyoming, South Dakota, Utah	$(OH)_4Al_4Si_8O_{20}\cdot nH_2O$ Ex: 90% montmorillonite, 10% quartz, feldspar, mica, etc.	High. Na is adsorbed ion, $pH = 8\text{--}10$	2100–2450°F	Very high, gel-forming	Very high	Flake size less than 0.00001 in.
Montmorillonite Class IB, Southern bentonite Source: Mississippi	$(OH)_4Al_4Si_8O_{20}\cdot nH_2O$ Ex: 85% montmorillonite, 15% quartz, limonite, etc.	High. Ca is adsorbed ion, $pH = 4\text{--}6.50$	1800°F +	Slight, little tendency to gel	Very high	Flake size less than 0.00001 in.
Kaolinite Class IV, fire clay Source: Illinois, Ohio	$(OH)_8Al_4Si_4O_{10}$ Ex: 60% kaolinite, 30% illite, 10% quartz, etc.	Very low	3000–3100°F	Very low, non-gel-forming	Low	Fire clays often ground; therefore may be relatively coarse, or ground to a flour
Illite Class III, grundite Source: Grundy, Illinois	$(OH)_4K_y(Al_4Fe_4Mg_4Mg_6)$ $(Si_{8-y}\cdot Al_y)O_{20}$	Moderate	2500°F ±	Low, non-gel-forming	Moderate	

*Adapted from R. E. Grim and F. L. Cuthbert[4].

The strength of the sand–clay mixture increases with increasing amounts of water, to a maximum point, after which the strength of the mixture decreases. The normal range of moisture in molding sands is compatible with the workability of the molding materials. Moist clay is thus the bond or binder of molding sands. The word *clay* is applied to a particular group of minerals which vary from fire-clay (kaolinite) to Western or Southern bentonite (montmorillonite) and a few special clays (halloysite and illite).

The bentonites are the most commonly used types. Table 4–1 lists some of their differences in chemical composition, atomic structure, base-exchange characteristics, swelling and shrinkage tendencies, and refractoriness.

Research into the chemical and physical character of clays has shown that both composition and structure have a great influence on the working properties of clays. The distinctive property of clay is its plasticity, which, in the presence of moisture, is imparted to a mass of sand.

The bonding forces involved in holding particles of clay together may be accounted for by several theories: electrostatic bonding, surface tension forces, and interparticle friction bond.

The mechanism of electrostatic bonding of clays may be described as a network of dipolar forces operating at the sand–clay and clay–clay interfaces. This network of forces is initiated by the preferential adsorption of positive ions and negative ions on combined water and clay (hydrated) surfaces. Figure 4–3 shows a micelle or hydrated clay particle.

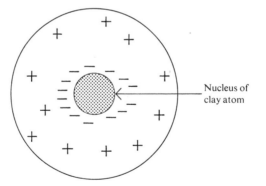

Fig. 4–3. A clay dipole. Surrounding the particle of clay are negatively charged hydroxyl ions positioned at varying distances from the particle. Outside this layer are positively charged ions (usually hydrogen ions) also located at varying distances from the particle; hence the term *double diffuse layer*. This layer is rigidly attached to the surface of the particle and is considered to behave as a solid. (Courtesy American Foundrymen's Society.)

When water is added to a dry clay, the negative ions are adsorbed on the nuclei of the clay atoms and form an integral part of the crystal. The positive ions are attracted by the negative ions, but repelled by the nuclei of the clay atoms, with the result that the positive ions take up equilibrium positions. The water forms neutralized clay micelles whose kinetic energy causes them to move toward one another. There is thus then a force of attraction between positive ions themselves and the nuclei of the clay particles (see Fig. 4–4). As the distance between the clay micelles increases, the force of attraction increases and that of repulsion decreases, resulting in a net intermicellular force.

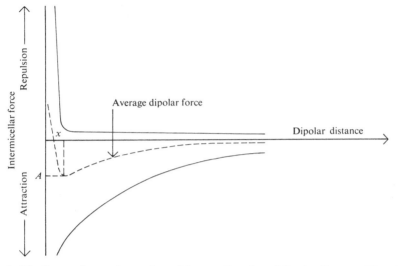

Fig. 4–4. The forces of attraction and repulsion as a function of dipolar distance. The critical dipolar spacing is defined as x units and the maximum bond is A. (Courtesy American Foundrymen's Society.)

The drawing together of two micelles causes the orientation of unlike ions, forming a clay dipole (Fig. 4–5), and a maximum attractive force is at an optimum distance of separation x. There are many such dipoles in a clay–water medium. Depending on the type of clay, a maximum degree of hydration is necessary to develop a dipole completely. This is why the strengths of clay-bonded sands increase with increasing amounts of water, up to a maximum value.

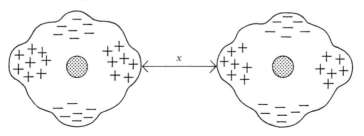

Fig. 4–5. Micellular dipoles, indicating the localized concentration of adsorbed negatively charged hydroxyl ions ($-$) and positively charged ions ($+$); x denotes critical intermicelluar spacings, the result of a compromise between the forces of attraction and repulsion.

As the amount of water is increased further, water enters the spaces between the dipoles to a distance greater than x (Fig. 4–6), resulting in a decrease in the net intermicellular force.

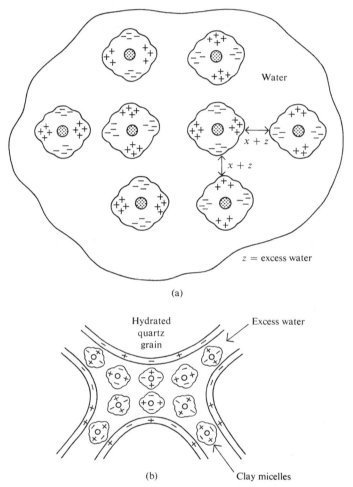

(a)

(b)

Fig. 4–6. (a) Dipole alignment of hydrated clay particles (condition of minimum free energy) in a water medium (green state). (b) Schematic sketch showing disposition of clay and quartz dipoles. In green sand the intermicellular voids are filled with water. (Courtesy American Foundrymen's Society.)

Surface tension of the water surrounding the clay and clay–sand particles provides another possible source of bond strength. The interstices of the clay particles are filled with water. The surface layers of water act on a stretched membrane of hydrated clay, forcing the clay particles together. When the amount of water is reduced by drying, the force holding the particles together increases.

The geometry of the aggregate can provide another force adding to the strength of the bond between particles. The theory of interparticle friction, or the so-called block-and-wedge theory, involves materials under pressure. When molding sand is rammed inside a flask, the particles are jammed against their neighbors. Sand which is rammed to a certain shape and which resists deformation is held together through interparticle friction. A favorable orientation of sand grains causes a packing action which supports the sand grains within the flask. When various sizes and shapes of sand grains are used, the strength properties of sand mixtures can be changed, which indicates the existence of an interlocking or frictional force.

Moisture in a molding sand is as essential as the clay substance itself. Moisture exists in two forms.

1) Free moisture which can be removed by drying.

2) Combined or absorbed moisture in the clay particles, which can be removed only by heating the clay to a high temperature.

Water which is present in amounts of about 1.5 to 8 % activates the clay in the sand, causing the aggregate to develop plasticity and strength. Water in molding sands is often referred to as *tempering* water. The water (up to a limited amount) is absorbed and held rigidly by the clay. Water in excess of that which can be absorbed by the clay exists as free water. Only that water held rigidly by the clay appears to be effective in developing strength. The rigid clay coatings of the sand grains may be forced together, causing a wedging action and thus developing strength. Free water can act as a lubricant, making the sand more plastic and moldable, though it may lower the strength. Thus it is evident that control of the water percentage in molding sand is very important.

In addition to the three basic ingredients of molding sand (sand, clay, and water), other materials may be present. They are usually added to improve certain properties, and are often referred to as *additives*. They are divided into two classes: organic and inorganic binders. The cereals, resins, proteins, pitch, and oils are organic binders. Cement, silicates, and some esters are inorganic binders.

A cereal additive, as used in the foundry, is finely ground corn, wheat, or rye flour, frequently gelatinized. Occasionally an organic additive is in the form of a fluid material, such as molasses. Cereals are used in the range of 0.25 to 2.00 % to increase the green or dry strength or the collapsibility of molding sands. Cereal binders are relatively inexpensive and impart good blowability and collapsibility to a core-sand mix. Cereals burn out at approximately 500 to 700°F. Thus they maintain their usefulness during molding and core-making, and do not disintegrate until they come into contact with the molten metal. Cereal binders, however, are hygroscopic in nature, and pick up moisture to such a degree that they can at times weaken the mold or core.

Resins and gums used as binders include a wide variety of both natural and synthetic compounds. Natural resins (gums) are derived from the exudations of plants or from distillation of wood, petroleum, etc. Synthetic resins, which are closely related to plastics, are the product of a reaction of several organic chemicals. The natural resins bake at about 400°F, somewhat more slowly than cereals, and may be used alone in amounts ranging from $1\frac{1}{2}$ to 8%. Such binders are often used when there is dielectric baking of the cores. Thermosetting resin binders require temperatures near 650°F to start thermal reaction. However, those resins contribute excellent strength properties, and the bake time required is relatively short.

Protein binders are of the casein type, casein being a calcined white powder derived from milk. Casein binders are water-soluble, and air-dry to a strong bond.

Pitch is a by-product of coke-making, being distilled from soft coals at about 350°F and above. Pitch is used in amounts up to 2.0% to improve the hot strength of the molding sand and the casting finish on ferrous castings.

Examples of binders are portland cement and the silicates. High-early-strength cement—mixed with sand in amounts varying from 8 to 12%, with from 4 to 6% water—is often used to make large molds and cores. The molds are generally air-dried for as much as 72 hours before assembly. Often it is necessary to air-dry them for a short period before removal of the pattern in order to maintain the shape of the mold wall. Cement-bonded sand develops high hardness and strength and does not burn out until about 2200°F, which makes possible sharp detail and close tolerance on large work. Cement-bonded sand does not absorb water after it has once been used, however, so it cannot be reused.

The inorganic compound sodium silicate (water glass) is being used more and more often, both in mold-making and core-making. If sodium silicate is mixed with a silica sand and exposed to air, it slowly sets or hardens on account of the reaction of carbon dioxide in the air. Hardening can also be accomplished in seconds if CO_2 gas is blown through the core or mold, and the molten metal poured immediately.

Sodium silicate binders develop high dry strength and do not easily burn out at normal temperatures, except when other additives to aid collapse are used. The process is commonly known as the CO_2 process.

Any material which does not promote binding action, but which is added to the sand in order to improve some special features, is called a *sand additive*. Examples of basic materials used as additives are sea coal, wood flour, silica flour, iron oxide, and dextrin.

Sea coal is a finely ground soft coal used in molding sands which are to be used for gray and malleable iron. Sea coal is added to the sand in order to improve the surface finish and the ease of cleaning of the castings. The seal coal should be finer than the molding sand to which it is added. It is used in amounts from 2.0 to 8.0%.

Many proprietary grades of ground-wood flour may be added to molding sands in amounts of 0.5 to 2.0% to enhance their normal stability. The wood flour functions as a cushion to control the expansion of the sand. It accomplishes this by burning out at elevated temperatures. One can readily see that if 5 to 8%

of combustible matter is present in the molding sand, a greater control of the expansion of the sand is possible. Cellulose, cereal hulls, and other combustibles are used for this purpose.

CONTROL OF GRAIN SIZE OF SAND

The term *sand-testing* is generally used to mean measuring certain physical properties of the mold and core materials prepared in sand plants. Silica sand is composed basically of the mineral quartz. A bonding agent such as clay must be added to it to hold the grains together. Silica sands or naturally bonded sands from various sources are tested for grain size and distribution. Naturally bonded sands often occur in nature; the silica grains are already mixed with a natural clay, and thus can be used directly as a molding material without adding any more clay. Figure 4–7 shows two methods of expressing the fineness of sand: (1) The size-frequency curve, in which the percentage retained on each sieve is plotted. (2) The cumulative curve points, which show the percentage of particles larger than the sieve size represented by a given point.

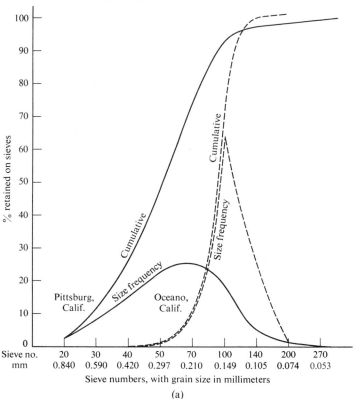

(a)

Fig. 4–7. (a) Curves constructed from the sieve analysis of two foundry sands show a comparison of size frequency and cumulative curves. (Figure continued on next page.)

U.S. series equivalent number	Amount of 50-gram sample retained on sieve		Multiplier	Product
	Grams	Percent		
6	None	0.0	3	0
12	None	0.0	5	0
20	None	0.0	10	0
30	None	0.0	20	0
40	0.20	0.4	30	12
50	0.65	1.3	40	52
70	1.20	2.4	50	120
100	2.25	4.5	70	315
140	8.55	17.1	100	1,710
200	11.05	22.1	140	3,094
270	10.90	21.8	200	4,360
Pan	9.30	18.6	300	5,580
Total	44.10	88.2		J5,243

$$\text{AFS grain fineness number} = \frac{\text{Total product}}{\text{Total percentage of retained grain}} = \frac{15,243}{88.2} = 173$$

(b)

Fig. 4–7. (*continued*) (b) Typical calculation of AFS grain fineness number. Size of sample: 50 grams; AFS clay content: 5.9 grams or 11.8%; sand grains: 44.1 grams or 88.2%.

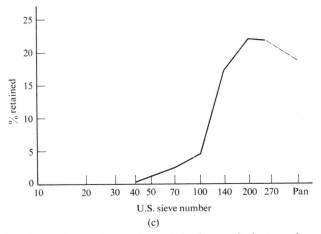

(c)

Fig. 4–7. (*continued*) (c) Graph of percentage retained on each sieve vs. sieve number; data taken from part (b). This sand has a high average fineness number, 173, and might be used for nonferrous castings. Usually the distribution curve looks more like a probability curve for most ferrous molding sands.

The advantage of a cumulative curve is that it shows a smooth curve, whereas the size-frequency curve may be a broken line connecting the different sieve points. If sand is purchased on specification, two cumulative curves can be drawn showing the limiting values for each sieve. Then, if the curve of the sand being tested falls between these two curves, it is satisfactory. Sieves that retain little material can be added without distorting the curve. If there are faulty sieves, when different samples of the same sand are sieved, the faulty places are shown by breaks each time at the same point in the curve.

Some combination of coarse and fine grains could produce rather rough castings and still show an apparently satisfactory average. The size of the voids between the grains, as well as the size and shape of the larger grains, should be checked. The presence of more than 10% of grains of No. 40 fineness (or 30, or 20) prevents the optimum finish from being obtained, since large grains are not conducive to smooth castings.

On the other hand, an excessive amount (more than 10%) of extremely fine grains, such as 200, 270, or finer, is not necessarily conducive to castings with smooth surfaces. A total of 10% or less of these fine-screen grains helps fill the voids and produces a smooth surface on the casting. The fineness of a given sand should be evaluated by the amount of it retained on each screen rather than the average retained on all screens.

CONTROL OF DISTRIBUTION OF GRAIN SIZES

The distribution of sand grains affects the appearance and degree of perfection of the casting. In addition, the amount of clay needed for control of the sand affects the expansion property of a given mold. A mold needs to be able to expand, if rupture of the mold surface is to be prevented. When there is maximum distribution of density, with three sizes—such as 60% coarse, 40% fine, and 0% medium grains—there may be scabs and buckles on the casting surface. Heavy additions of fine sand to close up a system, or heavy additions of coarse sand to open up a system may also result in scabs and buckles (see Chapter 13). When the sand grains expand, they have no room to move under these conditions. Thus a distribution of grain sizes is needed to provide optimum density, combined with self-cushioning. A sand with such a distribution is usually referred to as a *four-screen-spread sand*, that is, it is made up of sand in which 10% plus is retained on each of four sieves.

MEASURING THE COMPRESSIVE STRENGTH OF SAND

Green compressive strength may be defined as the maximum load required to break a green-sand mold under axial compressive stress. The load is expressed in pounds per square inch, and is applied on a specimen 2 in. in diameter and 2 in. high, having a standard rammed density.

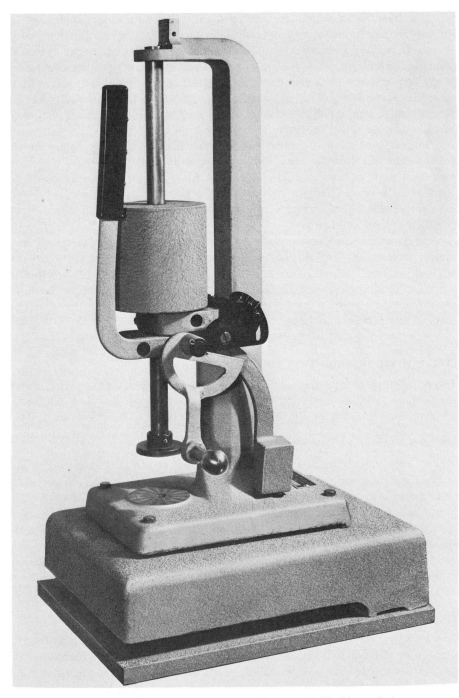

Fig. 4–8. A standard sand rammer. (Courtesy H. W. Dietert Co.)

The specimen is prepared by using a standard rammer and specimen tube (Fig. 4–8). The AFS standard specimen is prepared by ramming a known weight of sand with three blows by allowing a 14-lb weight to fall a distance of 2 in. The exact weight of sand necessary may be found by trial and error, or by using a density attachment. A synthetic molding sand bonded with bentonite may need a lesser amount of sand, while certain naturally bonded sands may need a greater amount to prepare a standard specimen. A stripping post (or pedestal) is used to remove the specimen from the tube; the specimen is then made ready for testing.

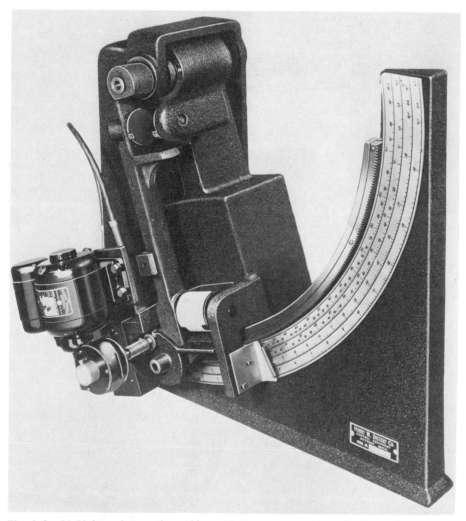

Fig. 4–9. (a) Universal strength machine with the proper specimen holder making a green shear test. [Parts (b) and (c) of this figure are on the following two pages.]

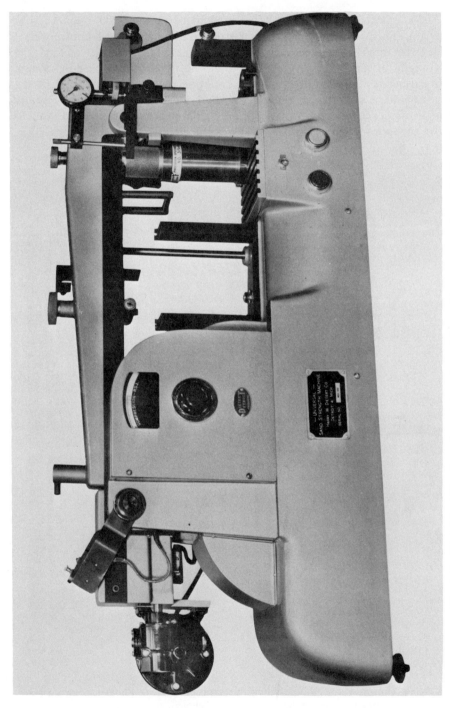

Fig. 4-9. (b) A No. 405 universal sand strength machine making a compression and deformation test.

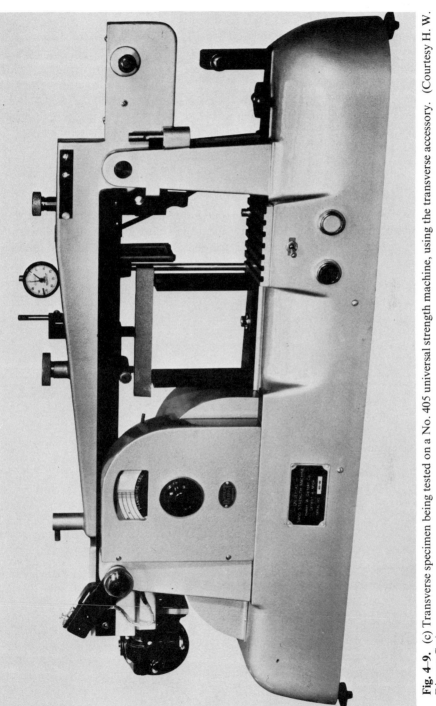

Fig. 4-9. (c) Transverse specimen being tested on a No. 405 universal strength machine, using the transverse accessory.   (Courtesy H. W. Dietert Co.)

The *universal strength machine* is used to perform compressive, shear, tensile, and transverse tests on dry or tempered sands. Figure 4–9 shows two models. The specimen is crushed between two compression heads which are designed to be self-aligning. A shear specimen of the same size is held in a positive shear holder until it fractures along its longitudinal axis. Compression and shear test results are in lb/in². The transverse specimen is 1 in. × 1 in. × 8 in. and is broken between two knife edges 6 in. apart, as shown in Fig. 4–9(c). The tensile specimen has a cross section of 1 in.² at its center.

GREEN SHEAR STRENGTH

The *green shear strength* of sand is the maximum shear stress which a tempered sand mixture is capable of developing. It is usually greatest at or near a moisture content of best workability, or correct temper. For a naturally bonded molding sand (used to prepare the chart in Fig. 4–10), this is 7% moisture.

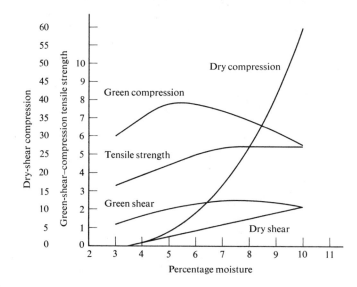

Fig. 4–10. The effect of moisture on natural plate-molding sand. Shear and compression are expressed in pounds per square inch, tensile strength in ounces per square inch.

The shear test may be made on the standard test specimen, after it has been removed from the specimen tube. The test is carried out by uniformly applying a load to the diametrically opposite halves of the two plane surfaces of the specimen, as in Fig. 4–9(a), at a rate of 24 ± 5 psi/min. The load is applied along the axis of the standard specimen.

GREEN AND DRY TENSILE STRENGTH

The tensile strength of a green molding sand is the maximum tensile stress which the sand is capable of sustaining when prepared and tested according to standard procedures.   A modified specimen tube for ramming is used with the universal-

Fig. 4–11. Accessory for measuring the tensile strength of core sand; part of the No. 405 universal strength tester. (Courtesy H. W. Dietert Co.)

sand-strength tests. *Green tensile strength* is the tensile strength of a foundry-sand mixture in the moist or tempered condition. *Dry tensile strength* is the tensile strength of a foundry-sand mixture which has been dried, such as a core-sand mixture. Measuring it requires an accessory such as that illustrated in Fig. 4–11.

PERMEABILITY

Permeability is that property of molding sand which enables air or gas to escape through the sand. Permeability is measured by noting the rate of flow of air under a known pressure through a standard AFS specimen while it is confined in the specimen tube. Standard permeability is determined by measuring the time necessary for 2000 cm³ of air to pass through the standard specimen while it is confined in the specimen tube under a pressure of 10 grams per cm². The permeability number may be calculated from the formula:

$$P = \frac{V \times H}{pAT}$$

where V = value of air = 2000 cm³,

$\quad\quad H$ = height of sand specimen in cm = 2.0 in. × 2.54 cm/in. = 5.08 cm,

$\quad\quad p$ = pressure of the air in grams per cm² = 10 g/cm²,

$\quad\quad A$ = cross-sectional area of sand specimen in cm² = 1 in² × 2.54 cm²/in² = 20.268 cm²,

$\quad\quad T$ = time in sec for 2000 cm³ air to pass through specimen.

The permeability number P is reduced to $3007.2/T$ sec.

A permeability test can be performed quickly and without calculation by using a unit such as that illustrated in Fig. 4–12. In this type, air under constant pressure (10 grams per cm² of water) is caused to flow through a selected orifice into the specimen tube and through the sand specimen. Another type of permeability-measuring device operates on a manometer principle. A reading is taken on a calibrated-sector scale which is rotated until it intersects the meniscus of the manometer water column. The permeability number is then read directly. Two different orifices may be used, 0.5 and 1.5 mm in diameter. The smaller is used for permeability of up to 49 ml/min and the larger for over 39 ml/min.

MOLD HARDNESS

Mold surface hardness, as determined by this test, is the resistance offered by the surface of a green-sand mold to penetration by a loaded plunger.

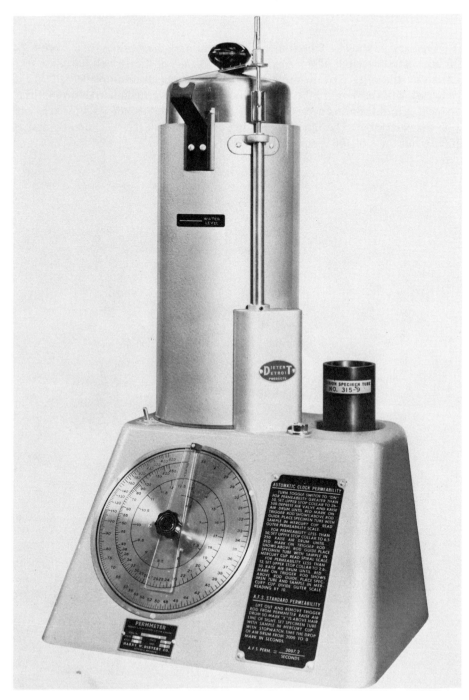

Fig. 4–12. One type of permmeter, or permeability-measuring device. (Courtesy H. W. Dietert Co.)

Figure 4–13 shows an instrument for determining the hardness of a green-sand surface. The principle of this test is similar to that of the Brinell hardness test: The softer the surface of the mold, the greater the penetration of the ball into it. Penetration of the point of the ball is indicated on a reverse dial, in thousandths of an inch. A mold offering no resistance to the ball has a zero reading. A mold having a hardness capable of completely resisting penetration of the ball has a reading of 100. A hardness of 100 is equivalent to infinity.

Fig. 4–13. Device for measuring the hardness of a green-sand mold. (Courtesy H. W. Dietert Co.)

Determinations of the hardness of a green-sand mold should be in the following ranges.

Type of mold	Hardness
Very soft rammed mold	20–40
Soft rammed mold	40–50
Medium rammed mold	50–70
Hard rammed mold	70–85
Very hard rammed mold	85–100

The practical advantage of using the mold-hardness tester lies in the fact that by this device one can standardize the degree of ramming. Hardness may be regulated by proper adjustment of the molding machine and the molding process. The maximum hardness to which a mold may be rammed can be determined by the mold-hardness tester. There may be a variation in hardness in different parts of the same mold; this indicates that there are variations in sand properties.

Increases in hardness of the mold improve the finish of castings, give more accurate dimensions of castings, and reduce penetration, drops, and swells. Excessive hardness of a mold may cause cracks, scabs, blows, and pinholes. Hardness of a green mold should be selected to match the sand, type of casting, and type and condition of metal.

CLAY

Clay minerals mixed with sand (about 2 to 50% clay) and a suitable amount of water give the molding sand strength and plasticity. Clay is thus the bond or binder of molding sands.

For testing purposes, the AFS clay in a molding sand is defined as those particles which fail to settle 1 inch per minute when suspended in water. These particles are usually less than 20 microns, or 0.0008 in., in diameter. This definition includes all very fine material, fine silica or silt, as well as the clay minerals present. The total percentage of these particles is called the *AFS clay content* of the sand.

Determination of the clay content is carried out as follows: A 50-gram sample of dried sand is placed in a wash bottle and washed with a solution of 475 ml of distilled water and 25 ml of caustic soda for five minutes; the mixture is stirred with a mechanical stirrer. It is allowed to settle for 10 minutes, and then 5 in. of water is siphoned off (Fig. 4–14). Fresh water is added, allowed to stand for 5 minutes, and again 5 in. of water is siphoned off. This continues at 5-minute intervals until the water is clear. The sand that remains is then dried and weighed. The loss in weight of the original 50-gram sample, multiplied by 2, is the AFS clay percentage in the sand.

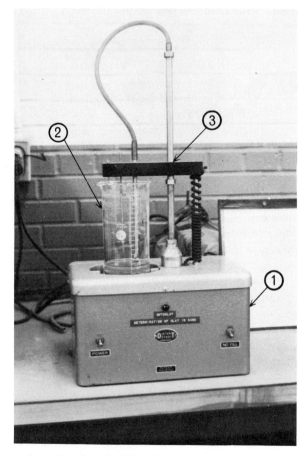

Fig. 4–14. Automatic sand washer. Sand is allowed to settle in water in beaker. The water is siphoned off, at 5-minute intervals, until water is clear. The machine fills and siphons automatically. (1) Control box for automatic siphoning and adding of water at 5-minute intervals. (2) A glass 1000-ml beaker. (3) Device that siphons water and adds new water during process.

MOLDABILITY INDEX

The *moldability-index test* is one which measures the amount of temper water required to produce a good molding mixture, over and above the amount of water required by the components of the molding mixture. A weighed sample of sand is riddled through a rotating cylindrical screen (Fig. 4–15). The weight of the molding sand passed through the screen, divided by 2 is the moldability index.

A 100 index indicates a dry-sand mixture containing additives but no temper water. The best moldability is obviously not 100. Each foundry must determine its own moldability level. The best moldability index for an average gray-iron foundry is usually between 65 and 80. For steel foundries it is between 40 and 60.

To ensure proper moldability, a constant moldability index should be maintained.

Fig. 4–15. Moldability machine for determining moldability of green molding sand. (Courtesy H. W. Dietert Co.)

COMPACTABILITY

Another useful test is the *compactability test*, which measures the percentage decrease in height of a riddled mass of sand, 100 mm high, under the influence of the compacting forces. It is an excellent measure of degree of temper of a sand, and provides a reference for determining the water requirements of any sand. The test is volumetric in nature, and is independent of the specific gravity of the sand. This enables one to compare mixes that differ greatly in the nature of their grains, such as carbon sands and zircon sands.

DILATOMETER

A *dilatometer* (Fig. 4–16) is used to study the behavior of molding sands at high temperatures. The dilatometer is equipped with an electric furnace and measuring devices necessary for evaluating sand samples being heated. One can test the properties of the sand mixture at various temperatures by setting the dilatometer at the temperature range desired.

Naturally it is very helpful to be able to test foundry-sand mixtures at elevated temperatures, for the results enable one to predict the mixture's behavior under the conditions that exist at the mold's pouring station.

BIBLIOGRAPHY

1. *Molding Methods and Materials*, first edition, American Foundrymen's Society, Des Plaines, Ill., 1962

2. *Foundry Sand Handbook*, seventh edition, American Foundrymen's Society, Des Plaines, Ill., 1963

3. Sarkar, A. D., *Mould and Core Material for the Steel Foundry*, Pergamon Press, London, 1967

4. Grim, R. E., and F. L. Cuthbert, *The Bonding Action of Clays*, part 1, University of Illinois Press, Urbana, Ill., 1945

5. Heine, R. W., C. R. Loper, and P. C. Rosenthal, *Principles of Metal Casting*, second edition, McGraw-Hill, New York, 1967

Fig. 4–16. One type of dilatometer for high-temperature sand tests. Oven is lowered over specimen tube and the temperature is raised to the desired range. A compression load is applied to the specimen and its breaking strength under the elevated temperature is recorded on the large dial indicator. (Courtesy H. W. Dietert Co.)

CHAPTER 5

MECHANIZATION OF FOUNDRY OPERATIONS

In the production of castings, as much as several hundred tons of material, varying with size and type of castings produced, may have to be continually handled. Thus if a large quantity of castings is needed, it is desirable to produce the molds required on a molding machine. Mechanization greatly reduces the time required to make molds.

The size and type of casting, and the volume of production and quantity of reproductions have a bearing on the degree of mechanization that is economically feasible. Mechanization can run from a simple operation in a small, low-production foundry to a completely automatic operation in specialized foundries, or in foundries which concentrate on repetitive production.

MECHANIZED MOLD-MAKING

Mechanization of molding operations is accomplished by using molding machines which are designed to perform the operations of compacting the molding sand and removing the pattern plate. Four common methods of compacting the sand around the pattern to make a mold are: jolting, squeezing, blowing and slinging (Fig. 5–1). The jolting and squeezing mechanisms are usually incorporated into one machine, though it is possible to have a separate machine for each operation in a production line.

Squeeze machines compact the sand by applying pressure on top of the mold, usually by means of high-pressure air acting on an air piston. Jolting is accomplished by raising the table supporting the mold and allowing it to drop against the machine base. With each impact, the grains of sand are compacted by the forces of deceleration acting on them. The jolting operation also requires the use of high-pressure air.

In the squeezing operation, the ramming of the sand is greatest at the sand–platen interface, or at the side from which pressure is applied. In the jolting operation, the ramming of the sand is at its highest level at the point at which the sand and pattern come together—the sand–pattern interface. To some degree, the ramming varies with the height (or drop) and the depth of sand in the flask. If the depth of the sand is not too great, both parts of the mold can be formed at the same time. When both these two operations are combined, the resulting machine is known as a *jolt-squeeze molding machine*.

Characteristics	Hardness Isofirms
Hand ramming 1. Produces molds of variable hardness 2. Is laborious and slow 3. Has low initial cost 4. Human equations high	
Jolt ramming 1. Mold lifted repeatedly and dropped 2. Is hard on equipment 3. Produces molds of uneven density 4. Is best for horizontal surfaces	
Squeezing 1. Suitable for relatively small work only 2. Is best for shallow flasks	
Sand slinging 1. Operation is fast 2. Has high initial cost 3. Ramming is uniform	

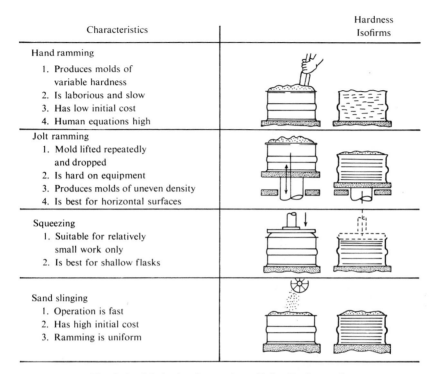

Fig. 5–1. Methods of ramming. (After Buchanan.)

Most of the molds made by machines are small enough so that the operator can lift and handle them. Molds that are too large and heavy to be lifted easily require a crane or hoist for this purpose.

A mold too large to be rolled over by hand, yet produced on a machine, can be rolled over directly by the machine (Fig. 5–2). The mold is compacted by jolting the table to which the flask is clamped. The table is raised and the mold is rolled over the center column of the machine onto the table behind the column. Sand is added to the cope side, and compacted by squeezing. The cope is then lifted and the pattern stripped automatically. The cope section is placed back onto the drag section, and the flask is removed. The completed mold is rolled onto a conveyor that transports it to the pouring area. One can also carry out this operation by making the drag half of the mold on one machine and the cope half on a companion machine, then assembling the molds at the conveyor area.

Molding machines can easily be used in cases in which large production runs are necessary. Reducing the number of slow hand operations and doing away with the need for skilled artisans results in a greater uniformity of molds and castings, as well as faster production and fewer defects. However, in order to justify the greater capital outlay involved in acquiring molding machines, there must be continuous high production runs. In addition, owing to their size and the complexity of the job they do, these machines are somewhat limited.

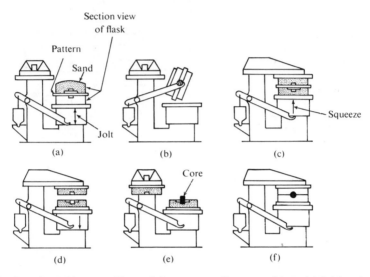

Fig. 5–2. Steps in making a mold on a jolt-squeeze-rollover machine. (a) Jolting the sand on the pattern in the inverted drag. (b) Rolling over the flask to fill the cope. (c) Squeezing to pack the sand in the cope. (d) Raising the cope to remove the pattern. (e) Placing a core. (f) Replacing the cope on the completed mold.

A sizable part of foundry operations involves the molding sand and the core sand, which are used in large quantities. The sand must be stored in clean, dry areas which are located near the places in which they are to be used. Compressed air is often used to transport dry sand from delivery cars to storage areas. Steel piping (with easy bends to permit the unrestricted flow of sand) is used to transport the sand, which is rendered fluid by air under pressure.

The preparation of sand for molding and coremaking requires a mechanical mixer to ensure complete mixing of sand with selected binders, special additives, and water. The mulling of the sand takes place in a machine such as that illustrated in Fig. 5–3.

The prepared sand is distributed to the molding stations by means of a belt conveyor system from the mulling station. Sand is diverted to overhead hoppers or storage bins (Fig. 5–4). Electromechanical controls at a central control station ensure that the proper amount of sand is distributed to each molding station as it is needed.

Whatever system is used for the conditioning of sand, the producing and assembling of molds, the melting and pouring of melt, all these factors have a pronounced effect on the quality of castings obtained.

Fig. 5–3. Typical slow-speed muller for mixing molding sand. (Courtesy National Engineering Co.)

Thus, if sand preparation and molding machines are combined with conveyors, a pouring station, cooling station, and shakeout area, the operations from sand mixing to shakeout may be mechanized.

The automatic-diaphragm technique is a variation of squeeze molding. Air pressure on a diaphragm is used to squeeze sand into the mold. A measured amount of sand is placed in the flask and the diaphragm unit is swung into place. Air is pumped into the chambers behind the pad, and at the same time the mold is squeezed by raising the pattern plate hydraulically; this makes possible uniform mold hardness, even in deep pockets.

Since the diaphragm technique results in uniform hardness of molds, precision engine-block castings are made in green sand. Less than 0.5% (by weight) variation has been reported in castings weighing more than 200 pounds.

Squeeze feet arranged strategically over the mold and operated by either pneumatic or hydraulic devices may also be used to apply selective pressure in any area of a mold that requires extra sand compaction.

One of the advanced methods of producing accurate castings with clean, smooth surfaces is the casting of ferrous and nonferrous alloys in air-dried sand molds, made by squeezing the sand under high air pressure.

The machine can be an automatic press or a manually controlled press. Its rating should be from 40 to 140 tons of squeezing force. When molds are prepared by this method, their surface finish is a function of the squeezing force. Pressures range from 327 to 2844 psi. Molds squeezed at pressures more than 1422 psi have markedly reduced roughness of the casting surface, as well as less mechanical burn-in. When pressures are raised to more than 2333 psi, any rounded sand grain in contact with the pattern starts to break down into a quartz flour, which coats the mold surface. The covering film of quartz flour on the surface of these air-dried squeezed molds combines with their feature of retaining a coarse-grained basic structure to ensure excellent gas permeability.

One machine consists of mold set-on and pick-off stations, high-pressure molding machine, rollover unit, and conveying equipment. It makes 140 molds per hour in a flask size of about 31 × 43 × 10/10 in. However, it is rather large, requiring about 15,000 square feet of floor space.

A flaskless molding machine that produces up to 750 molds per hour with twin-track operation is perhaps revolutionary in casting techniques. In this all-automatic molding machine, molds are made in a molding chamber; their vertical end surfaces contain impressions of the pattern on both sides. They are pushed forward on chutes which are a continuation of the molding chamber. Thus two rectilinear strings of molds are formed (Fig. 5–5). (See page 102.)

The end surfaces of the molds carry impressions of the pattern so that a mold cavity forms between each mold when two are booked together. No flasks are used. Molds which are approximately 14 × 18 in., and which are up to 8 in. thick, are produced by blowing sand from an overhead hopper into a boxlike metal mold cavity, which acts like a four-sided flask, except that the mold is on edge, like a slice of bread, rather than flat, as in conventional molding. At the front and back vertical faces, patterns are mounted on hydraulic rams; the cope is on one surface of the mold and the drag on the other (Fig. 5–6).

After the mold has been blown at a pressure of 90 to 100 psi, the patterns are squeezed on the vertical surface front and rear, at pressures up to 2000 psi. The mold is pushed out onto a track and the operation is repeated.

The molds then in line are poured on a continuous basis (Fig. 5–7). After

Fig. 5-4. Prepared sand being distributed to the molding stations by means of an overhead belt conveyor system.

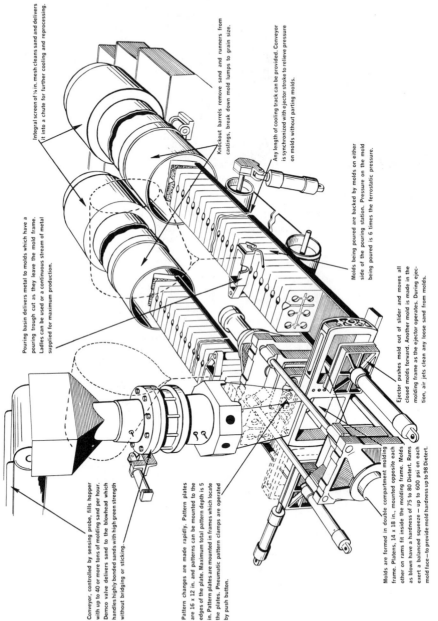

Integral screen of ⅛ in. mesh cleans sand and delivers it into a chute for further cooling and reprocessing.

Knockout barrels remove sand and runners from castings, break down mold lumps to grain size.

Any length of cooling track can be provided. Conveyor is synchronized with ejector stroke to relieve pressure on molds without parting molds.

Pouring basin delivers metal to molds which have a pouring trough cut as they leave the mold frame. Ladles can be used or a continuous stream of metal supplied for maximum production.

Molds being poured are backed by molds on either side of the pouring station. Pressure on the mold being poured is 6 times the ferrostatic pressure.

Ejector pushes mold out of slider and moves all closed molds forward. Another mold is made in the molding frame as the ejector operates. During ejection, air jets clean any loose sand from molds.

Conveyor, controlled by sensing probe, fills hopper with up to 40 or more tons of molding sand per hour. Demco valve delivers sand to the blowhead which handles highly bonded sands with high green strength without bridging or sticking.

Pattern changes are made rapidly. Pattern plates are 16 x 12 in. and patterns can be mounted to the edges of the plate. Maximum total pattern depth is 5 in. Pattern plates are mounted in frames which locate the plates. Pneumatic pattern clamps are operated by push button.

Molds are formed in double compartment molding frame. Platens, 14 x 18 in., mounted opposite each other on rams fit inside the molding frame. Molds as blown have a hardness of 75 to 80 Dietert. Rams exert a balanced squeeze — up to 600 psi on each mold face — to provide mold hardness up to 98 Dietert.

Fig. 5–5. Flaskless molding machine. (Courtesy Automated Molding Systems, Inc.)

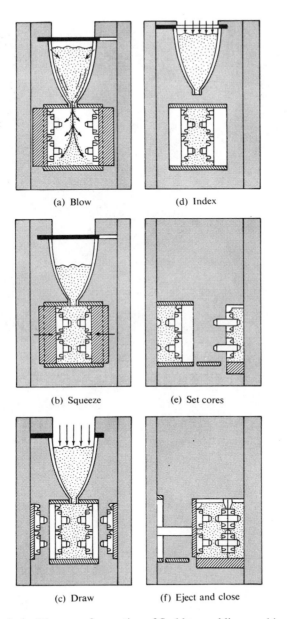

(a) Blow   (d) Index

(b) Squeeze   (e) Set cores

(c) Draw   (f) Eject and close

Fig. 5–6. Diagram of operation of flaskless molding machine.

cooling, the sand is recycled and reconditioned and the castings are carried on conveyors to cleaning stations. The castings have excellent finish and maintain a tolerance of ± 0.005 in. across the parting line.

Fig. 5–7. Two single-track flaskless molding machines with molds being poured.   (Courtesy Diesamatic Co.)

MECHANIZED CORE-MAKING

Cores are made both by hand and by machine. Many of the machines used for core-making are similar to those used for making molds.

Many cores are made by means of jolt-ramming on a jolt table. After jolting, the balance of the sand may be hand-rammed (for smaller cores) or pneumatically rammed (for large cores). Squeezing of the cores may be incorporated with the jolting. The simple jolt machine is quite versatile in the size and shape of the core boxes that it can ram.

The versatile Sand Slinger, so useful in mold-making, embodies the principle of ramming up large core boxes (Fig. 5–8). A large quantity of sand is delivered in a short time to the core box. The slinger method makes possible a more uniform packing density.

Fig. 5–8. Sand slinger being used for a small floor molding flask. (Courtesy Beardsly Piper Co.)

The most common core-making machine is the *core blower*, which provides the means by which small- and medium-sized cores may be rapidly produced. The principle of its operation is an air stream which carries core sand. It simultaneously fills the core box and rams the sand. The machine usually has a movable reservoir of sand, from which sand is blown into the core box. Between the reservoir and the core box there is a steel plate with holes that are adapted to the shape of the core box. When air pressure is applied, the sand is blown through the holes in the steel plate, and fills the core box. There is usually a sealing gasket between the plate and the core box, plus a means of clamping the two together, so that sand will not blow out at the joint between them. Vents are placed in the core box so that the air can be released to the atmosphere. The blowing action is very rapid; often the core box is filled in only 1 or 2 seconds. Air pressure is important if a core is to be blown to maximum density. Pressure should be of the order of 90 to 110 psi. The opening of the blow plate and the maximum clamping facilities determine the limiting size of the core box.

Core blowers are available in various sizes, from a simple hand-operated bench-type blower to a very large blower that produces cores for automobile engine cylinders, automatically, by the hot-box process.

The principle of *shell molding*, as presented in Chapter 3, is also used in making shell cores. The core box is attached to two heating plates. When these plates are clamped together, they present a heated chamber in which the thermosetting resin binder can react (Fig. 5–9). The unit is inverted so that the resin-coated sand can fill the core box. In the sand chamber, air under pressure ensures the uniform filling of the core box. After a short dwell period, the machine is returned to its normal position, and the unheated sand drains back into the chamber. The sand shell is allowed to cure for a short time before the core box is opened and the core removed. The shell may be from $\frac{1}{8}$ to $\frac{1}{2}$ in. thick, depending on the size of the core. The cores have excellent detail, smoothness of surface, and a close dimensional tolerance of ± 0.005. The cores need no further work, and may be used almost immediately.

The shell core machines may either be hand operated, and produce a number of small cores, or completely automated, and produce large cores.

Stock cores are standard-size cores that are commonly used. Foundries often make supplies of these by extruding sand through a die with a stock-core machine. The core sand is pushed through the die with a worm, similar to that found in a meat grinder. Sizes range from $\frac{1}{2}$ to 2 inches in diameter. The sand must be properly mixed with an oil binder so that it will hold its shape.

The cores come out in a long piece, and may be cut to the desired length and tapered when they are to be used. Round, square, and even hexagonal shapes are made in this manner. Cores may be composed of sand that is oil bonded or that is bonded with a thermosetting resin. The latter is mostly used to make cores in the form of long shell tubes.

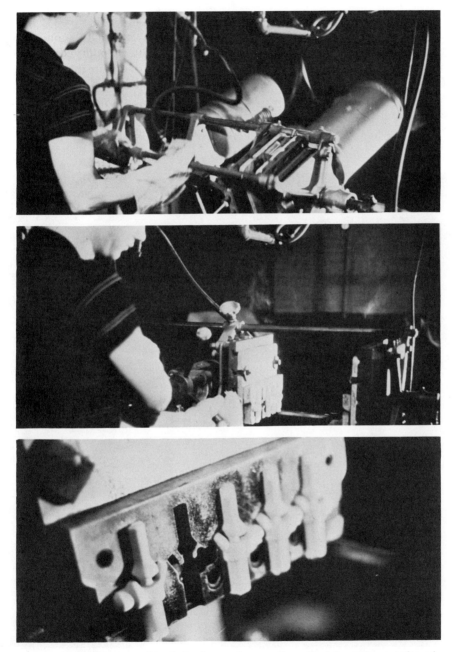

Fig. 5–9. (a) Shell-core machine being inverted manually so that the resin-coated sand can fill the core box. (b) Opening the two halves of the core box to remove cores. (c) Core box open and shell cores being removed by hand. (Courtesy Wentworth Institute.)

MECHANIZED HANDLING OF MATERIALS

A fully automated foundry system, in which the push of a button controls all aspects of casting production—from molding to shakeout, including core-making and core-setting—is already a reality. All these operations require the application of technology.

Foundrymen of the future will be faced with demands for product quality far beyond the past and present requirements. If foundrymen are to be able to meet the challenge, they must carefully plan for mechanization, often leading into automation.

Fig. 5–10. One of two sand mixers in a foundry. One is used for conditioning molding sand and the other for conditioning core sand. Sand is fed from the hopper to the sand-distributing system overhead. (Courtesy Electro Alloys Division, Abex Corp.)

Machines today are being designed and built to perform molding, core-making, sand-mixing, pouring, shakeout, and cleaning. When these machines are integrated with materials-handling equipment, foundries become mechanized.

About 4 or 5 tons of sand must be handled per ton of metal poured in a typical ferrous foundry. Therefore mechanization can well start at this point of operation. Sand must be mixed constantly, through the use of a muller or high-speed mixer (Fig. 5–10). But to adequately use a muller, one needs a conveyor system of either a moving belt or of air blowing the sand, to bring the sand to the muller, and then to carry the mixed sand on to the molding stations. In some cases a scoop truck is used to bring the sand to a skip hoist, which in turn discharges the sand with other materials into the muller. Sand, bentonite, water, and other additives are automatically metered into the muller. The prepared sand is discharged continuously to a conveyor belt, then plowed off the belt to hoppers at any of the molding stations.

MECHANIZATION OF THE MOLDING PROCESS

Perhaps the next consideration should be the mechanization of molding. Flask delivery lines are set up to feed cope and drag flasks to molding machines. Pattern plates are indexed into position, sand is fed into a flask, and a jolt-squeeze-draw cycle completes a mold-forming operation. In some cases a blow-squeeze molding machine is designed to produce a cope and drag simultaneously, in less than 12 seconds.

When copes and drags are produced simultaneously on two machines, they are passed to a rollover unit, where they are positioned, parting face up. At this point, cores may be set, either manually or automatically. Then the cope enters a rollover unit, and at the mold-closing station the drag is raised hydraulically into the cope. The closed molds are then passed through a clamping device and onto the conveyor, which carries the mold to the pouring station.

To derive the full benefit of mechanization, a foundry should be set up so that the molten metal is poured manually by ladles controlled from a moving line integrated with moving trains carrying molds. Or there should be an automated pouring system which is indexed to each mold car, engaging with it, and the metal should be poured from the bottom of the holding furnace or pouring box. This is a very quick cycle; in some cases it takes only 10 seconds. When this system is used, there is a direct reduction in metal spillage during pouring; the metal is also free from slag and dross, because it is poured from under the metal surface. However, this system permits only a narrow range of pouring temperatures.

To sum up the benefits derived from mechanization of molding: A foundry achieves increased production, requires less manpower, obtains a higher yield of castings, and a better quality of castings.

MECHANIZATION OF THE SHAKEOUT OF CASTINGS

In a mechanized situation, the poured molds travel through a cooling tunnel, after which they reach a shakeout station. An air-cylinder device is used to punch out the mold from the flask. In some cases, the conveyor cars are tipped, and the molds are allowed to slide onto a vibrating shakeout table. The sand is caught on a lower conveyor belt and brought back to the sand-mixing area.

The castings fall onto oscillating conveyors and are carried to cutoff machines, then on to grinders and again on to a skip hoist that feeds them into an airless blast cleaning machine.

MECHANIZATION OF THE METAL CHARGING

The charging of metal—whether into a cupola, an induction furnace, or another type of furnace—is another area in which mechanization has taken place. Mechanization provides fast charging and accurate composition of charge, and minimizes the handling of materials. An electromagnet is used to transfer iron from the scrap pile to the charging bucket. With nonferrous metals, a conveyor is used to carry the metal from bins to the charge platform and, in some cases, to run off the charge into the furnace.

When large amounts of loose bulky material are to be handled, pay-loaders provide a versatile service.

No matter what materials a foundry must handle—whether loose bulk or palletized—there is a need for constant study of the latest and best methods of handling materials. Pressurized tank trucks can deliver bulk dried sand, oils, resins, silicate, and alcohol to the foundry, thereby cutting the cost of bonding materials. Ingots of metal can be mechanically stacked and palletized for easy shipment to the industry. Fork-lift units can transport heavy materials to locations within the foundry. They are also useful when it comes to shipping finished castings.

Many foundries use pressurized air to transport sand and additives to the mulling station. Conveyor belts carry molding sand and core sand to the work stations. Conveyor cars, such as small trains, carry the molds to shakeout areas and then return the empty flasks and bottom boards to the molding stations.

By now it should be clear to any thoughtful reader that, in the fast-changing foundry industry, mechanization—of mold-making, of core-making, and of materials-handling—offers the solution to many problems.

BIBLIOGRAPHY

1. Heine, R. W., C. R. Loper, and P. C. Rosenthal, *Principles of Metal Casting,* second edition, McGraw-Hill, New York, 1967

2. Herman, R. H., "Sloan's New Automation Foundry," *Foundry* **91,** pages 52–61, April 1963

3. Miske, J., "Full Automation," *Foundry* **92,** pages 104–113, October 1964

4. Huskonen, W. D., "Mechanized System Cuts Cupola Charging Costs," *Foundry* **92,** pages 54–57, October 1964

5. Grundstrom, D. W., "Automated Molding," *Foundry* **94,** pages 72–75, April 1966

6. Huskonen, W. D., "Automated Brass Castings at Globe Valve," *Foundry* **96,** pages 58–63, March 1968

CHAPTER 6

THE STRUCTURE AND NATURE
OF CAST METALS

The study of the casting properties of alloys is closely allied to the study of the structure and properties of liquid metals, as well as the study of crystallization theories. Practical data and modern theories of physics and chemistry agree that liquid metals, and alloys near the melting point, have structures which resemble those of crystals of the solid metal.

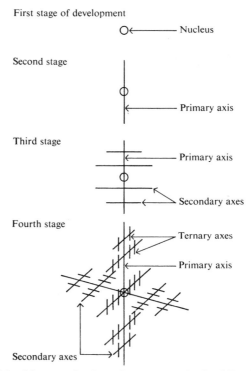

Fig. 6–1. Sketch of dendrite growth. Branches grow gradually, filling in to complete a solid mass of metal.

In a given body of molten metal, the coolest parts begin to solidify first. Microscopic crystallites of metal called nuclei form, and grow in a treelike fashion. This is called a *dendritic structure* (Fig. 6–1). The arms or branches continue to grow on one another until the mass of liquid metal is completely solid. In metallurgical terminology, dendrites are called *metal grains* or *crystals*.

This growth pattern is an ordered structure of many unit cells packed upon and around one another. Figure 6–2 shows three types of unit cells. Aluminum, copper, and austenitic iron are of the face-centered-cubic structure; ferritic iron and chromium are of the body-centered-cubic structure; and zinc and magnesium are of the hexagonal-close-packed structure.

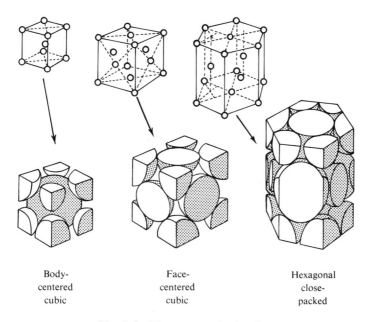

| Body-centered cubic | Face-centered cubic | Hexagonal close-packed |

Fig. 6–2. Three types of unit cells.

Solidification can be depicted graphically by cooling curves (Fig. 6–3). The solidification process is a function of time and temperature. When a pure metal changes from a liquid to a solid, the physical change takes place at a constant temperature level. An alloy in which the metals are in solution solidifies with a constant downward change in temperature.

Pure metals have certain characteristic properties that make them desirable for certain applications. Among these properties are high electrical and thermal conductivity, plus higher melting point, higher ductility, lower tensile and yield strengths, and in some cases better resistance to corrosion than is found in alloys. However, in the casting of pure metals, there are difficulties which affect the making of sound castings. Pure metals exhibit a tendency toward severe reactions when they are in molds; often there is cracking. Because of the way pure metals solidify, it is often hard to riser the casting properly to produce sound castings.

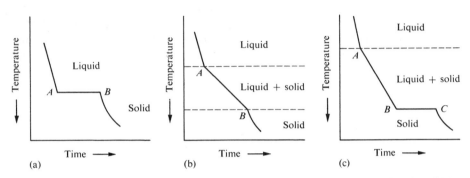

Fig. 6–3. Cooling curves, indicating the changes that take place during solidification of (a) a pure metal, (b) a solid solution, and (c) an insoluble alloy.

The solidification temperature and the melting temperature of pure metals are normally the same. Figure 6–4(a) shows a time-temperature cooling curve of a pure metal cooling through its solidification range.

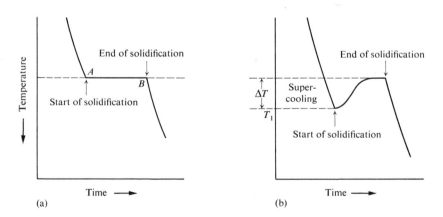

Fig. 6–4. Cooling curves of a solidifying pure metal. (a) Equilibrium cooling. (b) Cooling curve, showing supercooling.

Heat is dissipated by the molten metal at a fairly constant rate until the metal reaches the solidification temperature at A. The temperature then remains constant throughout the period of change of state from liquid to solid. The time, represented by the distance AB, corresponds to the time needed for the formation of nuclei, dendrites, and solid mass. The curve from point B down to room temperature is the curve typical of cooling bodies.

When pure metals are cooled rapidly, the formation of solid crystals is sometimes retarded. Below the solidification temperature, the metal demonstrates supercooling. At a temperature T_1 below the solidification temperature, solid crystals suddenly nucleate. The heat of fusion liberated at this point increases the temperature to that of the solidification point. Figure 6–4(b) illustrates this so-called supercooling. This phenomenon can also occur in alloys; when it does, it can affect the structure and properties of these alloys. An example of this is found in gray cast iron. When gray cast iron is cooled rapidly, the amount of graphite flakes is less than in a gray cast iron which has been cooled slowly. The size, shape, and orientation of the graphite flakes may also be different.

The combining of two metals in an alloying operation produces different behaviors, depending on which metals are involved. The metals may be completely soluble in the liquid state. Or they may be completely insoluble, and will form two layers, much as oil and water do. Such alloys remain separate when they solidify, and are known as *mechanical mixtures*. Some metals, such as tin and cadmium, are completely insoluble in one another in the solid state. Usually there is a limited solubility; aluminum–copper alloys are of this type.

PHASE DIAGRAMS

In alloys in which there is between 5.65 and 52.5% copper and the balance aluminum, there are two different kinds of crystals. One is a solid solution with a high percentage of aluminum and the other is a solid solution which contains a high percentage of copper. Mechanical mixtures usually solidify by a *eutectic reaction*. An example of this is the aluminum–copper phase diagram (Fig. 6–5), which shows a 10% copper, 90% aluminum alloy. If this alloy were used as a metal for casting in sand molds, it would begin to solidify at 1160°F (627°C). The first solid dendrites would consist of 2% copper and 98% aluminum. As cooling continued to just above 1018°F (550°C), the solid material would increase; it would have a composition of 5.65% copper and 94.35% aluminum. The remaining liquid material would be composed of 33% copper and 67% aluminum.

When the point of maximum solubility for copper was reached, no more copper atoms could be accepted into the aluminum lattice. Further solidification could occur only if new copper-rich crystals were to become available. This would occur at the *eutectic temperature* of 1018°F.

These copper-rich crystals are beta crystals, composed of 52.5% copper and the rest aluminum. At the same time, aluminum-rich crystals are formed. This mechanical mixture of fine copper-rich crystals and fine aluminum-rich crystals

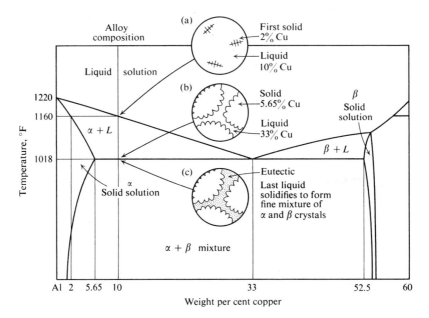

Fig. 6–5. Phase diagram of the aluminum-rich end of the aluminum-copper phase diagram Microstructures shown are obtained by very slow solidification.

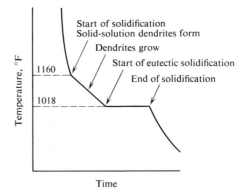

Fig. 6–6. Cooling curve of a eutectic alloy. The eutectic metal solidifies at a single temperature, much the same as pure metal does.

solidifies, forming what is known as a *eutectic alloy*. The eutectic alloy solidifies at a single temperature, and so forms bonds which are similar to those found in pure metal. Figure 6–6 diagrams the formation of a eutectic alloy produced by extremely slow cooling.

When metals solidify at a normal rate, as they do, for example, in sand or chilled molds, primary dendrites contain coring, which results from incomplete

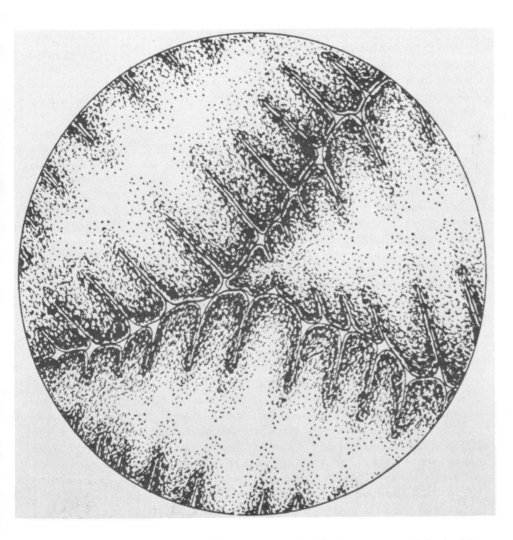

Fig. 6–7. Example of coring in a nickel-copper cast alloy. Darker areas toward the dendrite surfaces represent increasing copper content.

diffusion in the solid state during solidification. This fault is common during solidification of nonferrous alloys (Fig. 6–7). A reheating of the casting allows complete diffusion to take place. Thus, when the casting is allowed to cool slowly, heat treatment serves this purpose, and is not detrimental to the mechanical properties of the casting.

We have said that coring during the solidification process is quite common in cast alloys. What are some of the effects of coring? The microstructure of the final casting, as well as the temperature range over which an alloy freezes, is affected by coring. For example, in the case of a 50–50 copper–nickel alloy, the continued segregation of copper in the liquid metal eventually results in the formation of pools of liquid which contain substantially more than 50% copper. These pools have a lower final solidification temperature than that predicted by the phase diagram for the 50–50 alloy. Hence solidification of the cored alloy is not complete until a temperature somewhat below that of the equilibrium solidus is reached. In this case pools of copper have formed.

Alloy systems obey certain rules, whether they are liquid, solid, or gaseous. When at minimum energy at any temperature, they are *in equilibrium*. Equilibrium is said to be reached if the temperature, pressure, volume, and composition of each phase suffer no measurable change in a long time under fixed external conditions.

The number of phases that can be in equilibrium at atmospheric pressure is one more than the number of components. With a pure element, only two phases may exist in equilibrium: solid and liquid at the melting point. In binary (two-component) alloys, three phases may coexist at one temperature and atmospheric pressure: liquid, liquid plus solid, and solid. All phases must be at the same temperature. If not, heat would move from the hotter to the colder phases, and they would not be in equilibrium.

SOLID SOLUTION ALLOYS

Two metals which are mutually soluble in their liquid state behave much like water and alcohol. The liquid alloy looks like a one-substance liquid. Two metals in alloy appear much the same. This resemblance to a single substance may exist in solid alloys also. Such an alloy is said to be a *solid solution* because the two metals are mutually soluble in each other. Silver and gold are a pair of metals that are completely soluble in each other in both the solid and liquid states. Silver or gold may be added to one another in any proportion with little effect on the microscopic appearance of the solid alloy, except for a change in color. Copper and nickel behave the same way. When elements are added to form an alloy, the strength of the alloy is greater than the strength of the component elements.

If an alloy is composed of 50% copper atoms and 50% nickel atoms, the alloy will have a single-crystal structure that is face-centered-cubic, because both copper and nickel are face-centered-cubic metals. Figure 6–8 shows a phase diagram or equilibrium diagram for this system. The line labeled liquidus indicates that all above it is a liquid solution. This is called a *single solution* because both metals are mutually soluble at the elevated temperature. Below the line marked solidus, the metals are in *solid solution*. At temperatures between the liquidus and the solidus, two phases are present. This zone has both liquid and solid material present, and is often called the *mushy zone of solidification*.

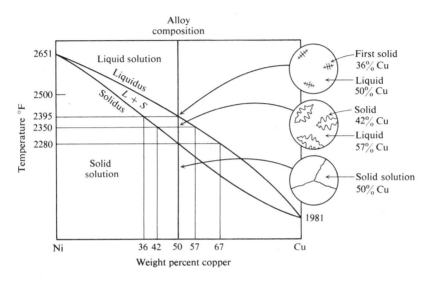

Fig. 6–8. Phase diagram for the nickel-copper alloy system. Microstructures shown are obtained by very slow solidification.

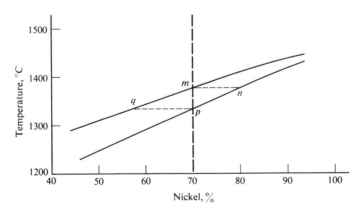

Fig. 6–9. A detail of the nickel-copper system. For the 70-30 alloy, *m* represents the temperature at which solidification begins, *p* the temperature at which the alloy is completely solid, *n* the composition of the first solid particle to form, and *q* the composition of the last trace of liquid as solidification nears completion.

A vertical line drawn at 70% nickel on the diagram of Fig. 6–9, shows that an alloy of 70% nickel begins to solidify at 1375°C (2500°F). When we draw the two horizontal lines *mn* and *qp* connecting the liquidus and solidus curves, we see that the first bit of metal to solidify contains about 80% nickel at point *n*. The high nickel content of the solid portion is at the expense of the liquid, which now has an average nickel content of less than 70%. With the lower content of nickel, the solidifying temperature of the liquid is lowered. As cooling continues, the percentage of solid increases and the percentage of liquid decreases. The nickel content of both the solid and liquid decreases until the end of solidification. The solid at *p* contains 70% nickel and at *q* there is about 56% nickel in the last traces of liquid.

The composition of the liquid portion of the alloy at any temperature in the freezing range is given by the upper line, and the composition of the solid portion by the lower line. At 2460°F (1350°C) there is 62% nickel in the liquid and 74% nickel in the solid. It can be said that the composition of the liquid follows the liquidus line from *m* to *q* and the composition of the solid follows the solidus line from *n* to *p*.

For every alloy composition, the solidus line indicates the temperature at which melting begins. The liquidus line indicates when melting is complete. Melting and solidification may be considered the reverse of each other. If the metal is allowed to solidify at a slow rate, the atoms will diffuse and produce a homogeneous alloy containing 70% nickel and 30% copper.

Phase diagrams are important devices which utilize graphs to show the temperature at which a change of some kind takes place in an alloy. They indicate the phase present at equilibrium in a given alloy at any temperature, and from this one can deduce the physical and mechanical properties of the alloy. Phase diagrams indicate solidification temperatures, and thus are useful in determining the required pouring temperatures of a given alloy. The solidification range gives us a clue to many foundry characteristics, such as shrinkage, hot tearing, and fluidity.

The diagrams in Fig. 6–10 give an approximate picture of the formation of two different forms of crystals for solid solutions on a copper basis, solidifying in a narrow and a wide range. From this we can see that the greater the solidification range of the copper-tin alloy, the greater the number of equiaxed grains or small crystals developed. Copper-aluminum alloys, which solidify in a narrow temperature range, yield castings in which the crystals have a pronounced columnar grain structure.

Thus equilibrium or constitutional diagrams of alloys form the very basis of metallography. These diagrams help metallurgists to choose alloys which have predictable casting properties; that is, alloys which have a given relative shrinkage, fluidity, susceptibility to hot-shortness, shrinkage, and porosity.

One method of determining variations in the casting properties of alloys is to construct *composition–casting-property curves*. Such curves disclose a systematic connection between the character of the curves and the process of solidification of alloys as determined by the positions of the alloys in the equilibrium diagrams.

Composition–strength,   composition–plasticity,   composition–temperature–strength, and similar curves are used to reveal the variations in the service (mechanical) properties of alloys. When one is assessing the technical properties of casting alloys, one should not give preference to any one property in particular, but rather one should consider the combination of several properties. For instance, among the many requirements which casting alloys must meet, that of low shrinkage is given prominence, because this factor supposedly determines the susceptibility of an alloy to hot-shortness. Actually, alloys of a given composition which show either low or high shrinkage may exhibit considerable hot-shortness, or on the contrary they may exhibit no cracks at all.

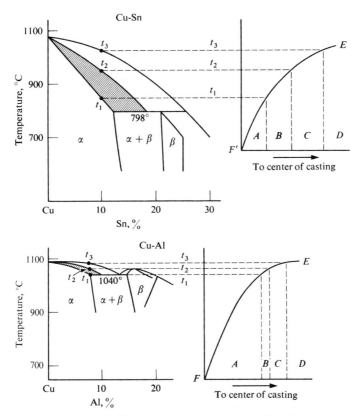

Fig. 6–10. Influence of the crystallization range on the width of liquid–solid zone in solidification. (*A*) Crystals; (*B*) crystals + liquid; (*C*) liquid + crystals; (*D*) liquid; (*FE*) curve of temperature distribution in casting.

The mode of solidification for different metal alloys, or the formation of crystals while the alloy is in a certain definite temperature range results in certain properties of the solidified metal. The temperature of the alloy at the time crystals are formed influences the properties of the cast alloy, and this plays an important part in the final characteristics of the cast metal. The systematic character of curves which depict the composition versus the volume of shrinkage pores and cavities is also found in equilibrium diagrams of alloys of the eutectic type.

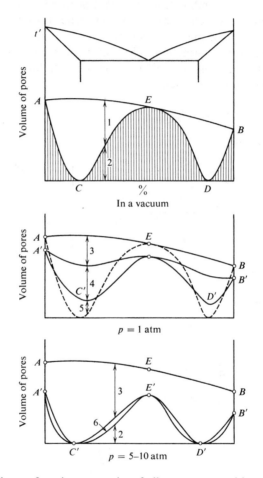

Fig. 6–11. Dependence of casting properties of alloys on composition and the form of the constitutional diagram: (1) Volume of dispersed porosity; (2) volume of pipes; (3) external shrinkage; (4) fine porosity; (5) shrinkage pipe; (6) volume of fine porosity.

Figure 6–11 shows such diagrams for the case of crystallization of alloys in a vacuum and under pressure. The curve AB, which indicates the possible maximum volume of shrinkage voids, shows only the total volume of the shrinkage voids; it does not show their character, or the way they are distributed in the metal. For both pure metals and eutectics, the variations in volume on solidification appear mainly as a concentrated shrinkage cavity or pipe at the hottest point of the casting. In alloys which solidify in a given temperature range, the volume of the concentrated shrinkage cavity will be less because the pores are dispersed. When certain alloys solidify under ordinary atmospheric pressure, there is impregnation of disperse (intradendritic) porosity, and for this reason there is external shrinkage, as well as a concentrated shrinkage porosity in such alloys. Such shrinkage does not take place when these alloys solidify in a vacuum.

When the atmospheric pressure above crystallizing castings is increased, the external shrinkage increases and the volume of disperse porosity decreases. This is true of alloys which solidify in a wide temperature range. Thus the composition–property curves show that the application of pressure above a casting is effective in eliminating porosity and increasing density of castings which solidify in a wide temperature range. This system does not always work for metals which solidify in a narrow temperature range.

Although these diagrams cannot provide an answer to every problem of foundry practice, they do help in giving a scientific basis for foundry practice.

THE SOLIDIFICATION PROCESS

It is during the process of solidification that the difference between gray, mottled, white, and chilled irons becomes apparent. The fundamentals of this process are brought out in the iron–carbon diagram, which describes graphically the effects that carbon has on iron, on its melting point, and also on the phases or constituents that form at different concentrations of carbon. Temperature influences these constituents as well as the way different carbon alloys solidify (see Fig. 6–12). A schematic representation of the iron–carbon diagram is given in Fig. 6–13.

In Fig. 6–13, the part indicated by *CEF* represents partial solubility. The portion *CE* is similar to a solubility curve. Above it, varying proportions of iron and carbon crystallize in the form of austenite. Metal solution that falls beneath the *CF* line begins with a low percentage of carbon. As the temperature lowers, the liquid becomes richer in carbon and the solidification temperature (line *CE*) drops. This continues until all the metal is solidified and the solid solution consists of austenite of varying carbon composition.

In molten metal that contains between 2.0 and 4.3% carbon, solidification proceeds as outlined above, except that the last remaining liquid solidifies as a eutectic. The eutectic is the last metal to solidify, and represents the composition at which both materials solidify at the same time.

The last free austenite to precipitate is austenite containing 2% carbon. The remaining liquid solidifies in the form of a eutectic consisting of austenite and cementite. The matrix consists of free austenite interspersed with the eutectic.

When molten metal is composed of 4.3% carbon, no precipitation begins until the eutectic temperature, 2066°F, is reached. The resulting solid is of a eutectic structure.

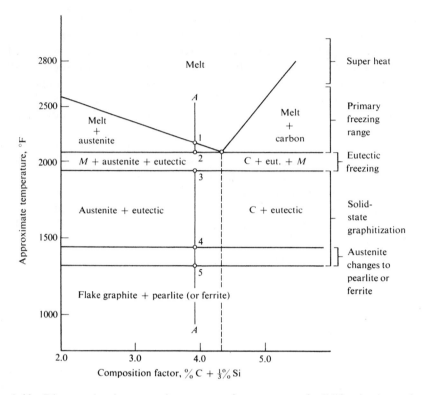

Fig. 6–12. Diagram showing approximate range of temperature of solidification in cast irons.

Iron with a carbon content greater than 2.0% is classified as cast iron. The carbon content of commercial irons range from 2.25 to 4.3%. These commercial irons usually have silicon and manganese to improve their properties. An iron whose composition is above 2.0% carbon in the eutectic has austenite interspersed with the eutectic. If this solid composition is cooled to 1380°F at a rate sufficient to retard the formation of graphite, a white cast iron is formed, consisting of pearlite, cementite, and ferrite (iron).

One property of iron—a property which of itself seems of little importance—is the ability to change from one atomic arrangement to another. At room temperature iron has the body-centered-cubic structure. But if it is heated, it changes (at 910°C) (1670°F) to the face-centered-cubic structure. If it is heated to still higher temperatures, it changes back (at 1390°C) (2534°F) to the body-centered-cubic structure. Such changes are called *allotropic transformations*. The production of steel tools, and the immense development of the manufacturing industry in the past 100 years are due, in part, to this transformation.

The photomicrographs shown in Figs. 6–14 through 6–23 (pages 126–134) illustrate the structure of iron, graphite, cementite, and pearlite.

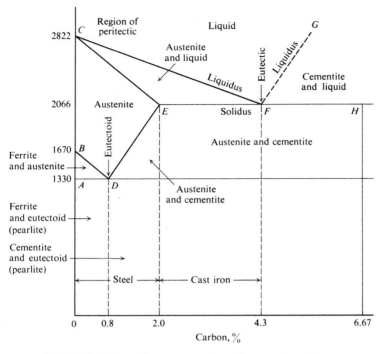

Fig. 6–13. Schematic representation of iron–carbon diagram.

BIBLIOGRAPHY

1. J. Datsko, *Material Properties and Manufacturing Processes*, John Wiley, New York, 1966

2. B. A. Rogers, *The Nature of Metals*, American Society for Metals, Cleveland, Ohio, 1964

3. American Society for Metals, *Metals Handbook*, 8th edition, Vol. 1, 1961

4. H. F. Taylor, M. C. Flemings, and J. Wolff, *Foundry Engineering*, John Wiley, New York, 1959

5. F. Weinberg and B. Chalmers, "Dendritic Growth in Lead," *Canadian J. Phys.* **29,** page 382, 1951

6. A. P. Gagnebin, *The Fundamentals of Iron and Steel Castings*, International Nickel Company, New York, 1957

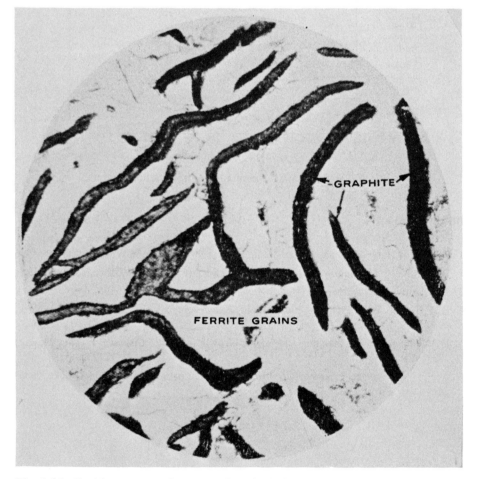

Fig. 6–14. Ferritic structure of gray cast iron (etched; 200 × approximately). (Courtesy A. P. Gagnebin–International Nickel Co.)

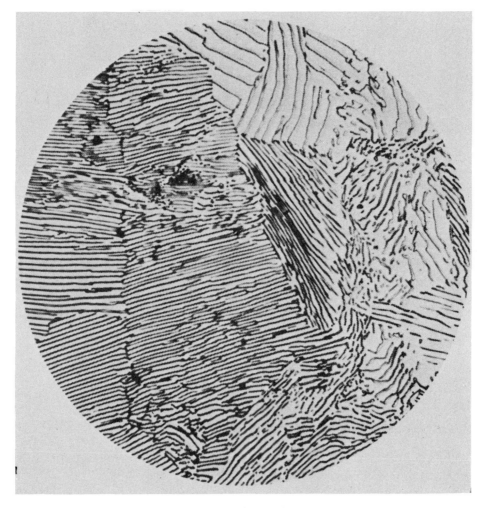

Fig. 6–15. Microstructure of pearlite (etched; $850\times$ approximately). (Vilella) (Courtesy A. P. Gagnebin–International Nickel Co.)

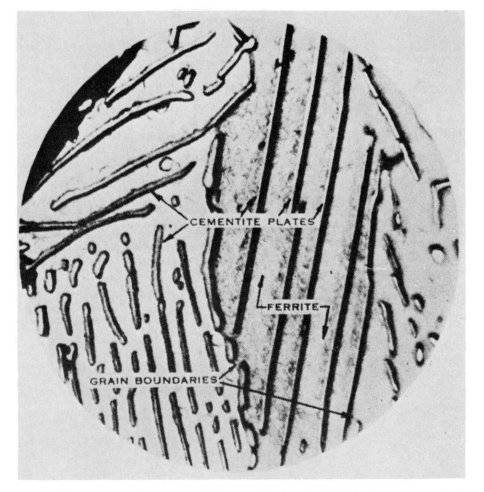

Fig. 6–16. Structure of pearlite at high magnification (etched, 2250× approximately). (Courtesy A. P. Gagnebin–International Nickel Co.)

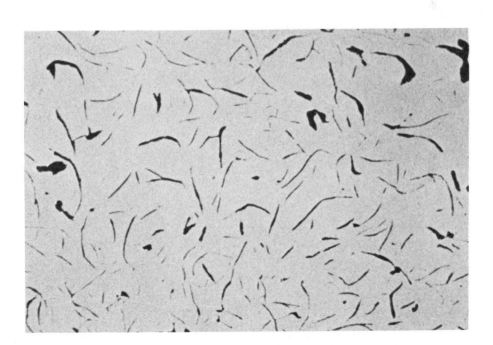

Fig. 6–17. Flake graphite in thin section of gray cast iron of approximately eutectic composition (unetched; 100× approximately). (Courtesy A. P. Gagnebin–International Nickel Co.)

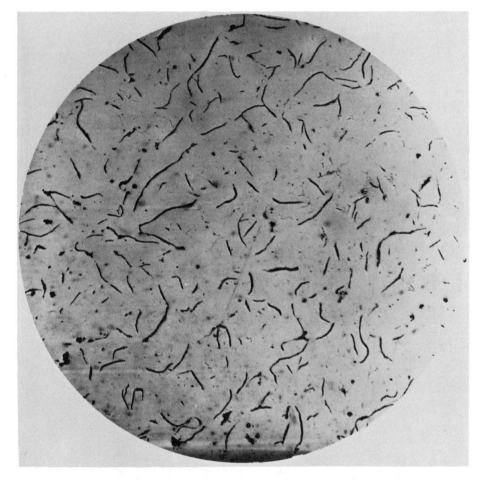

Fig. 6–18. Photomicrograph of medium-strength gray iron, Type A (65 ×). (Courtesy A. P. Gagnebin–International Nickel Co.)

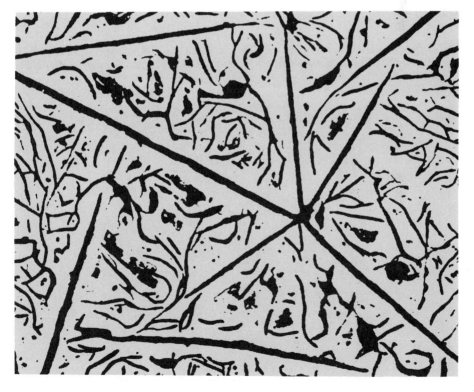

Fig. 6-19. Graphite in hypereutectic gray iron (unetched; 60 × approximately). (Hanneman and Schrader.) (Courtesy A. P. Gagnebin–International Nickel Co.)

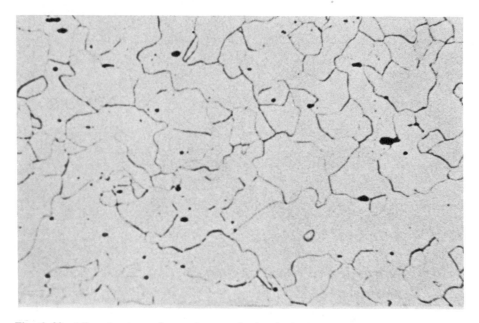

Fig. 6–20. Microstructure of pure iron or ferrite (etched; 100× approximately). (The British Cast Iron Research Association.) (Courtesy A. P. Gagnebin–International Nickel Co.)

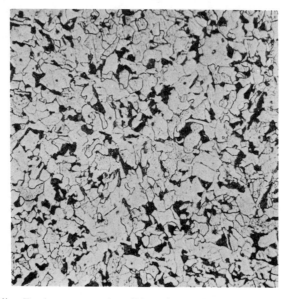

Fig. 6–21. Pearlite–Ferrite structure in mild steel containing 0.2% carbon (etched; 100× approximately). (Courtesy A. P. Gagnebin–International Nickel Co.)

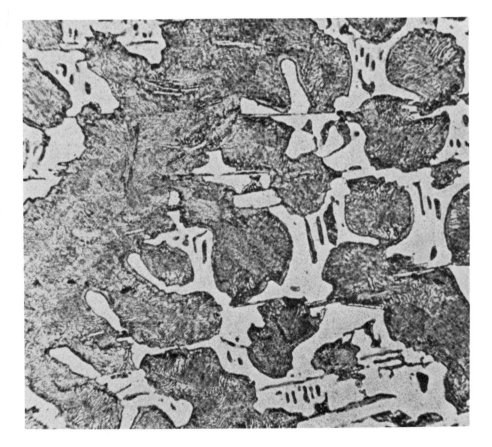

Fig. 6–22. Pearlite and iron carbide in white cast iron containing 2.9% carbon (etched; 250× approximately). (Courtesy A. P. Gagnebin–International Nickel Co.)

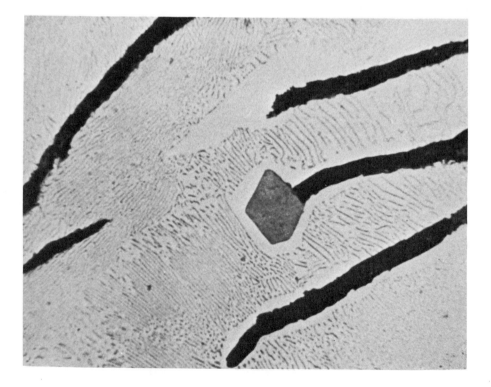

Fig. 6–23. Pearlite and graphite in gray cast iron (etched; 690 × approximately). (British Cast Iron Research Association.) (Courtesy A. P. Gagnebin–International Nickel Co.)

CHAPTER 7

SOLIDIFICATION OF METALS AND ALLOYS

FACTORS RELATING TO SOLIDIFICATION

Modern research techniques have shown that the solidification mechanism is directly related to the characteristics of pure metals and of metals in solid-solution alloys or eutectic alloys. During solidification, energy gradually leaves the liquid metal. To a casual observer, the process seems sudden, but if one is able to measure the change in energy, the process seems very gradual.

We can consider the general problem of heat transfer in relation to shaped castings in sand molds under three headings.

1. Basic laws of heat transfer

2. Variables of shape and size

3. Variables of mold and casting material

To apply heat-flow theory to the progress of solidification in sand-cast shapes, one needs to have available certain values of the thermal constants involved, since, for mathematical equations, one needs to assume certain constants.

Castings in the three basic shapes—plates, cylinders, and spheres—with volume-to-surface ratio and other factors being equal, solidify at different times. The solidification time of the three simple shapes can be calculated with reasonable accuracy. Castings such as plates exhibit minimum heat losses due to turbulence of 50°F, whereas naturally deep castings which are top-gated (such as spheres) exhibit maximum heat losses of 180°F (in the case of cast-iron spheres). Quietly poured shapes, such as bottom-poured cylinders, appear to have minimum heat losses on the order of those of plate castings.

The chilling power of green-sand molds varies considerably with ramming density, average grain size of the sand, amount of combustible materials in the sand, moisture used with the sand, and casting thickness.

Solid metals are crystalline in nature. Each crystal is made up of atoms, arranged in a definite pattern. Each metal has a different pattern, and in each the atoms are arranged in an orderly fashion. The spheres used in the laboratory to construct models of various lattices represent the volume or space in which an atom vibrates. If heat is applied to the metal, the "amplitude" of vibration, or the volume in which the atoms vibrate, gets larger. The spheres appear to get bigger. This is why metals expand when heated.

The way in which the casting solidifies influences the number of salable castings obtained. It also has a bearing on the interior soundness of the casting (or its interior defects). Therefore, questions concerning crystallization and the phenomena appearing in the course of solidification have always been of interest to foundrymen. An important phenomenon accompanying solidification is volume contraction, and the formation of shrinkage cavities connected with such contraction. Anyone who wants to produce a sound casting tries to minimize shrinkage cavities and shrinkage porosity in the casting.

A casting does not solidify suddenly throughout its entire mass. The thinnest sections, since they are most subject to the cooling influences of the surrounding mold, solidify first, and in so doing they undergo an internal contraction. To make up for this contraction, they draw liquid metal from adjacent heavier sections. These, in turn, when *they* come to solidify and contract, try to draw liquid metal from other adjacent parts of the casting. In a well-designed casting, this solidification and contraction proceeds progressively throughout the casting, the heavier sections feeding the lighter sections, until the last parts to solidify draw liquid metal from the risers. Good gating and running is also important, to ensure that, when pouring is complete, the coolest metal will be at the extremities of the mold and the hottest in the risers. Thus, when the casting cools, it does so progressively; the extremities solidify first and the risers last. This is called *directional* or *progressive solidification* (see Fig. 7–8).

When pure metals or alloys solidify and cool, the vibrations of their constituent atoms are reduced. Solid nuclei develop, enlarge, and thicken. The various branches or chains of atoms enlarge and grow, ultimately become all solid, and the metal contracts in volume. This is true of all metals, except some gray cast irons and bismuth. Cast metal contracts, or shrinks, in three distinct steps. As the metal cools from its pouring temperature to its solidification temperature it undergoes a liquid contraction. During the solidification period, solidification contraction takes place. As the solid metal cools to room temperature, solid contraction takes place.

Figure 7–1 shows these various changes in specific volume.

The total change in volume takes place in three stages.

1. Liquid shrinkage, which occurs as the liquid metal cools from pouring temperature to solidifying (freezing) temperature.

2. Shrinkage which occurs during the actual solidification or freezing.

3. Solid shrinkage, or contraction which occurs as the temperature of the solid casting cools from the solidification temperature to the ambient temperature.

It is the shrinkage which takes place during stages one and two, but particularly stage two, which causes shrinkage porosity in castings.

(One exception is in the case of gray iron. When graphite is precipitated out of the melt and solid structures, it occupies more volume than the pure iron, and thus causes expansion.)

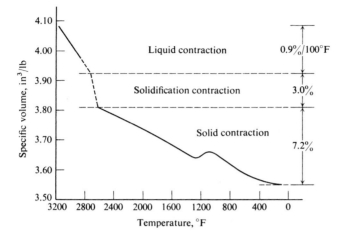

Fig. 7–1. The change in specific volume of solidifying and cooling steel.

An interpretation of contraction is illustrated in Fig. 7–2, for an imaginary casting which is cooled slowly so that no metal freezes to the sidewalls. The metal original level L_0 falls to L_1 as the metal cools to the solidification point, thus going through liquid contraction. The metal loses its superheat to the surrounding mold material. As the metal slowly and uniformly solidifies, there is a sudden solidification contraction, and the metal level drops to L_2. The liquid shrinkage for steel is approximately 0.9% per 100°F, and the solidification shrinkage is approximately 3%. As the solid casting continues to cool to room temperature, it continues to contract, and reduces to a size smaller than the mold (or pattern).

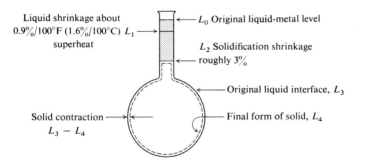

Fig. 7–2. Schematic representation of shrinkage.

The first two contractions—liquid and solidification—are directly related to the risering of a casting. If one is to produce sound castings, the solidification contraction is the most important to consider. The last or solid contraction results in the casting being slightly smaller than the mold cavity. That is why patterns are made slightly oversize with a patternmaker's rule to allow for solid contraction. Hot-tearing and internal stresses occur during the solidification contraction stage, and thus it is especially important to pay attention to this process.

When a pure metal is allowed to solidify in a mold, the portion of the liquid which is next to the mold wall reaches the freezing temperature first, and begins to solidify. It is in this area that heat extraction by the mold wall is the greatest, and therefore a thin skin or shell of solid metal forms against the mold wall, and surrounds the liquid center. As heat is extracted through this thin wall of metal, the liquid continues freezing onto it. Thus the wall increases in thickness. The solid metal progresses inward toward the center until the casting is completely solid. As the solid metal builds up, the liquid level drops. During solidification shrinkage, a so-called *shrinkage pipe* is formed in the thermal center of the casting or riser. A shrinkage pipe is a cavity formed by the contraction in the metal during solidification; it makes its appearance at the point where the last portion of liquid metal freezes, as in a riser.

Pure metals melt and freeze (or solidify) at the same temperature. Under equilibrium conditions, the metal is completely solid below this temperature. Cooling curves can be drawn to show that the metal cools quite rapidly to the freezing or solidification point, where the temperature remains constant while the metal loses its heat of fusion. Further cooling occurs only after the pure metal is completely solidified. Figure 7–3 shows a cooling curve of a pure metal.

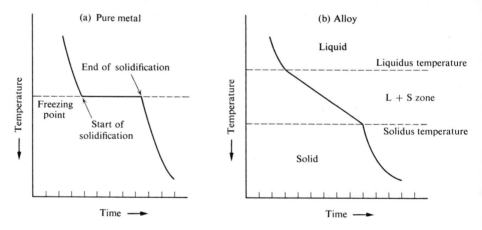

Fig. 7–3. Cooling curves of solidifying metal. (a) Equilibrium cooling of pure metal. (b) Ideal cooling curve of alloy.

It is desirable to know how rapidly a casting will solidify. The freezing rate has an important bearing on the degree to which one can eliminate solidification shrinkage and promote directional solidification. It also has a bearing on the

(a)

(b)

(c)

(d)

(e)

Fig. 7–4. Aluminum is poured into a cast iron mold, allowed to set for a certain number of seconds, and the mold is inverted to pour out the liquid metal. Note that the thickness of the wall increases with increasing dwell time. (a) A 10-second dwell time. (b) A 15-second dwell time. (c) A 25-second dwell time. (d) A 31-second dwell time. (e) A 37-second dwell time.

segregation (explained below) and the size of the grain formed. In a sand mold, the rate at which the mold accepts heat determines, to some extent, the solidification time. Figure 7–4 illustrates how a solid metal wall increases in thickness during freezing. It shows a cone of aluminum which was poured into cast-iron molds at room temperature. The molds were open at the top, and after the initial pouring, the remaining liquid was poured out at stated intervals.

Segregation of elements in the final casting takes place during solidification. It is usually on a microscopic scale, and involves the metal which is entrapped by the various branches of the dendrites. Each dendrite exhibits a segregation pattern; as a result, a single grain (the dendrite) differs in chemical composition from point to point.

The thickness of the metal skin solidified in any given time can be expressed as

$$D = k\sqrt{t} - c,$$

where k and c = constants, t = time, and D = thickness of metal skin.

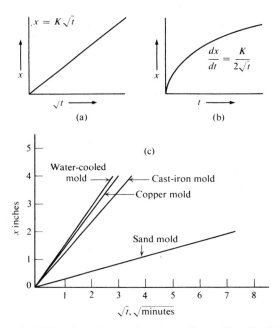

Fig. 7–5. (a) Rate of solidification of a metal against a flat mold wall. (b) Thickness of solid metal, x, plotted as function of time t. (c) Solidification of steel in molds made of different materials.

SOLIDIFICATION RATE

In metal that freezes against a large flat wall which has a normal pattern of heat flow to the mold surface, the thickness of the solid metal is proportional to the square root of the time, or

$$x = K\sqrt{t},$$

where x = thickness, t = time, and K = constant.

Figure 7–5 shows the parabolic relationship between thickness and time. Freezing begins at $t = 0$. The initial freezing rate is very fast, since $dx/dt = K/2\sqrt{t}$.

On a microscale, solidification progresses by the thickening of the trunks and branches of the dendritic skeleton, as shown schematically in Fig. 7–6. Solidification progresses because heat flows from the metal to the mold. Therefore the coldest metal is at the interface between metal and mold, and the surface is the first region of the casting to develop solid material. Thus the growth of solid metal from the mold surface to the interior is of a gradual, progressive nature.

During the progressive solidification of an alloy from the mold wall inward, there may be (1) a completely liquefied zone, (2) a mushy zone consisting of liquid

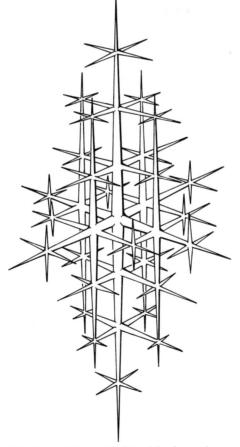

Fig. 7–6. Schematic of dendrite formation.

metal plus dendrites, and (3) a completely solid zone. The nature of the development, and the location of the mushy regions in a casting, are most important to its ultimate soundness.

The schematic drawing in Fig. 7–7 illustrates a practical case of solidification with differential and interacting growth next to the mold wall, which results from the presence of corners and risers. Near the external corners the greater volume of sand removes heat from the metal more quickly, causing the solidification to proceed at a relatively fast rate. The smaller volume of sand at internal corners removes heat less quickly, and causes solidification to proceed at a relatively slow rate.

Progressive solidification, as it continues, takes on wavelike aspects. There is a lateral wave growth which may be thought of as occurring when waves travel toward the interior of a casting, or when *freeze waves* start and finish (Fig. 7–8). When they start, the crests of the freeze waves travel with the growing dendrites into the free liquid; this marks the beginning of solidification. The troughs of the freeze waves travel with the growing dendrites until the troughs meet, and the last remaining liquid solidifies. In between the waves is the intermixed liquid and solid metal, which is the zone of solidification.

Variables arise out of the differences between the thermal properties of the mold and the solidifying metal. A mold material which has high heat capacity and thermal conductivity induces a high rate of progressive solidification. A sand mold which has a low heat capacity induces a slow rate of progressive solidification.

When the design of the mold is such that this progressive solidification is carried on through the casting to an appendage known as a riser, this is called directional solidification. The technique of directional solidification is used to encourage all the shrinkage to take place in the riser. With this technique, solidification starts in one part of the mold and gradually moves in a desired direction; it does *not* simply start in any area in which molten metal is needed to feed the casting.

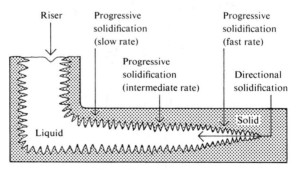

Interacting wall growth

Fig. 7–7. Schematic of wall growth.

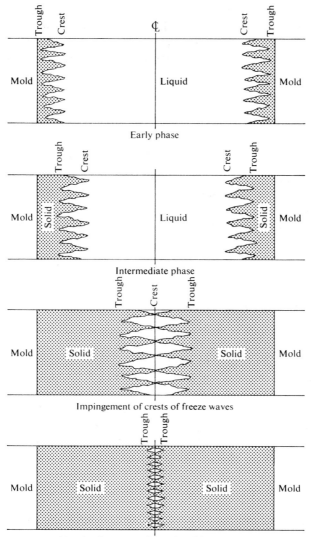

Fig. 7–8. Mechanisms of lateral wall growth by progression of crests and troughs of freeze waves.

An alloy which has a narrow solidification range also has a narrow mushy zone. Steep temperature gradients have the same effect, in that they narrow the mushy zone. One can increase the steepness of thermal gradients by chilling the casting (pouring the metal against a metal mold wall which is cold). The width of the mushy zone is important because it is this factor that determines the grain structure. Narrow mushy zones give rise to a columnar crystal structure, whereas wide mushy zones may result in the formation of an equiaxed crystal structure. The following variables affect the extent of the mushy zone.

a) *Solidification range.* The wider the solidification range, the wider the mushy zone.

b) *Thermal gradient.* The steeper the thermal gradient, the narrower the mushy zone.

c) *Types of alloy* and *degree of solid solubility* of elements.

The temperature gradient is itself dependent on freezing (or solidification) temperature and thermal conductivity of the cast alloy and the mold material. High freezing temperatures and low thermal conductivities give rise to steep thermal gradients. Chilling the mold naturally steepens the temperature gradients.

SOLIDIFICATION AT JUNCTIONS

Junctions such as L and T junctions must be given special consideration. Because a junction is normally heavier than any of the sections which it joins, it usually cools more slowly than adjacent sections.

The method of inscribed circles (shown in Fig. 7–9) can be used to predict the location of hot spots, which are sites of final solidification and possible shrinkage. In the L-shaped section, the largest circle which can be drawn in the junction is larger than the largest circles that can be drawn in the walls. The same is true of a T-shaped section, in which the circle at the junction is even larger than the one for the L-shaped section. The larger circles in both the junctions are places in which there may be hot spots, which will turn out to be unsound unless special precautions are taken.

Figure 7–10 shows the progress of solidification, as indicated by the shaded areas. Hot spots are indicated by the light areas within the large circle inscribed at the junction.

The joining of two walls may result in an L-, V-, X-, or T-shaped junction. If small fillets and rounded corners are used in the L- or V-type junction, a heavy section is formed. The radii used should be such that the thickness of the junction is the same as that of the adjoining walls (see Fig. 7–11). The area within the dashed lines shows the amount of metal which should be eliminated to avoid hot spots. The thickness of the wall at the junction can of course be reduced even further by using radii such that the junction will be thinner than the adjoining sections.

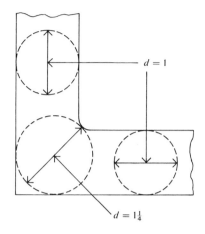

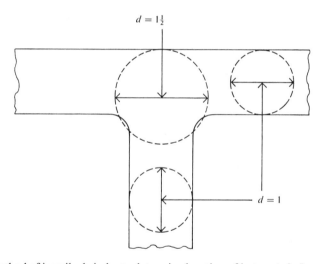

Fig. 7–9. Method of inscribed circles to determine location of hot spots in L and T junctions.

An X-shaped section has a still greater tendency toward hot spots and un-soundness than L- or V-shaped sections. One way out of this dilemma is to use a hollow core (Fig. 7–12) to reduce the wall section. A better method is to stagger the sections so as to produce T-shaped junctions, which can be more easily controlled.

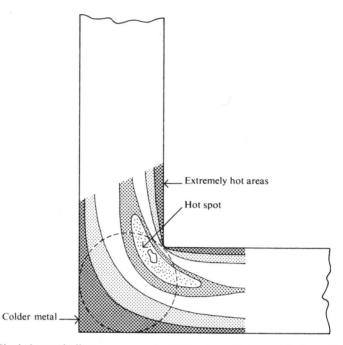

Fig. 7–10. Shaded areas indicate progress of solidification. Hot spot is indicated by irregular unshaded area that falls within the inscribed circle.

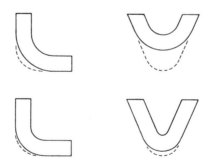

Fig. 7–11. Reduction of cross section in L- and V-shaped sections.

SOLIDIFICATION TIME

To predict the solidification times of castings which are geometrically similar but of different size, one can use Chvorinov's rule, as follows. When molten metal or alloy is poured into molds which are made of the same material and which are at the same temperature, the time required for solidification of the cast shape is

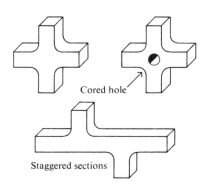

Cored hole

Staggered sections

Fig. 7–12. Reduction of cross section in an X-junction.

proportional to the square of the volume of the casting divided by the square of the surface area of the casting. Thus a 2-in. steel cube freezes in one-fourth the time required for a 4-in. steel cube.

By applying Chvorinov's rule, one can compare the solidification times for castings of varying shapes:

$$\text{Time} = K_2 \, (\text{volume/surface area})^2.$$

SOLIDIFICATION OF ALLOYS

In pure metals, the interface between the solid skin and the liquid metal is a nearly smooth wall. When small amounts of alloying elements are present, they tend to be rejected by the solidifying metal at the liquid-solid interface. These added elements usually lower the melting point of the pure metal, and alter the mechanism of solidification. Freezing occurs with dendrites projecting out into the liquid metal. Thus the liquid-solid interface becomes jagged as minute pine-tree-like crystals protrude into the liquid metal. In metal that is almost pure, dendrites do not project very far. At the late stages of solidification, the dendrites reach across the remaining liquid and interlock.

The first stage in the solidification of a casting is the formation of a thin chill layer, owing to the fact that the temperature of the mold is considerably lower than that of the metal. The chill layer or zone usually has a fine-grained structure caused by the rapid cooling and large number of nuclei which form. Dendrites start to grow from the inner boundary of the chill layer toward the thermal center of the mold.

The resultant crystals are necessarily elongated or columnar under usual conditions, for dendritic growth is preferential in the direction of the temperature gradient. Their direction thus indicates the direction in which a casting solidifies. As the growing columnar crystals approach the interior of the casting, the thermal gradient is diminished by the dissipation of superheat. This makes it possible for other crystals to nucleate at random rather than having to continue growth in columnar grains. Figure 7–13 illustrates this process schematically.

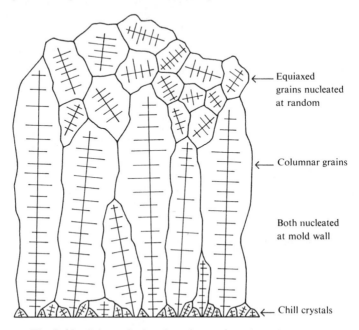

Fig. 7–13. Schematic drawing of a cross section of a casting.

During later solidification in the central region, where liquid and solid remain, more grains form, which are equiaxed like those at the mold surface, but larger. Metals that freeze with a wide mushy zone have all-equiaxed structures.

Most foundry alloys freeze over a fairly wide temperature range, and with flatter temperature gradients than pure metals.

Some cast iron, aluminum, and magnesium alloys—as well as most brasses and bronzes—freeze in sand molds without forming any appreciable solid skin. Often in aluminum castings the central part of the casting solidifies even before the surface is completely frozen.

SOLIDIFICATION SHRINKAGE OF ALLOYS

Metals that freeze with a wide mushy zone have liquid and solid metal mixed throughout. If the liquid pools are not interconnected, solidification shrinkage can take place in isolated or random spots throughout. If the liquid areas are kept open, then one can feed the spots more molten metal through an external source such as a riser and prevent solidification shrinkage (often in the form of micro-shrinkage). Localized shrinkage is found where the last liquid freezes, but often the liquid entrapped between dendrites also shrinks, developing widely distributed voids referred to as *microporosity*.

Commercial metals freeze over a range of temperatures that can be predicted from the phase diagram for the particular metal (see Fig. 7–14). Because of thermal gradients, which are active in the solidification process near a mold wall, three specific zones exist:

1. Completely solid zone

2. Zone composed of solid dendrites and liquid metal

3. Completely liquid zone

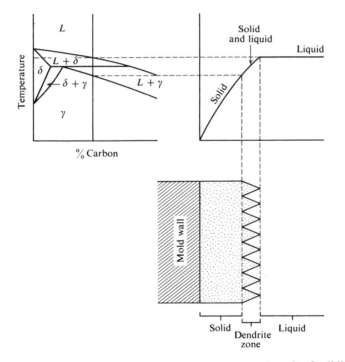

Fig. 7–14. Relationship between iron-carbon phase diagram and mode of solidification of a 0.30% carbon steel casting.

Two, or sometimes even three, of these zones often exist concurrently (Fig. 7–8). They may be visualized as two waves traveling inward from the mold-metal interface. The tips of the dendrites projecting into the liquid are the places at which the freezing starts and the base of the dendrites, where the last remaining liquid solidifies, is the place at which the freezing ends. In between these two points is a mixture of liquid and solid metal called the mushy zone. This is the zone in which actual solidification occurs.

In some gray irons in which freezing begins with the formation of dendrites which grow into the liquid metal, provision must be made to feed the casting with liquid metal to prevent shrinkage during solidification. Because of the high carbon level of most cast irons, however, the quantity of these dendrites is limited. They do not form at all in soft eutectic irons. Thus cast iron does not freeze by progressive growth of an envelope of solid metal.

Those few dendrites that do form in cast iron grow out into liquid iron which has a eutectic composition. The *eutectic* form of a metal is that alloy which has the lowest melting point possible for the given composition. Eutectic solidification takes place at many centers or nuclei, which grow to form so-called *eutectic cells*. During dendritic solidification, there is a decrease in weight of a given volume of metal, resulting in solidification shrinkage. When eutectic solidification begins, the growth in size of the graphite (carbon) within the eutectic cells results in an expansion in the volume of the metal rather than the shrinkage which takes place during the freezing of most pure metals. Graphite has one-third of the density of iron. In other words, a given weight of graphite occupies roughly three times the volume of a similar weight of iron.

The low density of the graphite precipitated during eutectic solidification causes considerable expansion, which is often enough to prevent formation of shrinkage defects at this time. In some cases, few or no risers are needed under these conditions. When a volumetric change does become evident, it is attributed to either a solidification shrinkage or to dilation of the mold cavity. Dilation of the exterior of the casting, or movement of the mold wall, is due to expansion pressure because of the formation of graphite while the interior metal is still partly liquid. This dilation of the exterior of the casting is primarily due to the fact that the rejection of graphite causes an expansion of the alloy during freezing. The expansion produces dimensional variations of the castings.

GASES IN CAST METALS

Even under the best melting conditions, it is frequently impossible to prevent gas —particularly hydrogen—from dissolving in metal in its molten state. This hydrogen comes mainly from the moisture in the air, which breaks down into hydrogen and oxygen at the elevated temperature of the metal. When the metal solidifies, the hydrogen is liberated, causing many small voids, called *pinholes*, and some large holes, called *blow holes*, in the castings. These holes are of course objectionable, and should be eliminated or kept to a minimum.

Any holes in a casting, whether large or small, tend to be detrimental to mechanical properties. Thus one must use great care to see that the molten metal is free from dissolved gases before it is poured into molds.

Hydrogen is the most common offender, and is soluble in almost all metals to varying degrees. Figure 7–15 illustrates the behavior of hydrogen in various metals. The solubility of hydrogen also gives us an idea of the solubility of other gases. Any gas becomes quite soluble at high temperatures. This solubility decreases as the temperature of the molten metal decreases to solidification temperatures, at which point the greatest portion of gas-caused porosity develops within the casting.

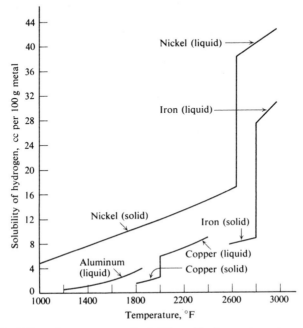

Fig. 7–15. Effect of temperature of solubility of hydrogen in various metals.

The method of overcoming pinhole porosity varies with different metals. For most metals, hydrogen porosity can be reduced by observing the following rules.

1. Keep the temperature of the metal as low as possible consistent with the pouring temperature.

2. Hold the metal in the molten condition no longer than necessary.

3. Keep all items that come in contact with the metal *dry*.

4. Agitate or stir the metal as little as possible. Do not disturb the surface of the molten metal any more than necessary.

5. Skim the metal, preferably when ready to pour.

The presence of hydrogen is especially serious when the casting must be pressure-tight, or when the product is to be polished, plated, or anodized. Because it is virtually impossible to prevent gas pickup, before the casting is poured one must treat the melt with a degassing agent such as an inert gas like chlorine, nitrogen, or argon, or one must use a solid degasser as a source of the inert gas.

When used in conjunction with a proper flux, a degasser speedily removes hydrogen and other harmful gases. Introduced into the melt, it evolves chlorine and volatile chloride compounds that bubble through the molten metal as scavenging gases, sweeping out hydrogen, oxides, and all nonmetallic inclusions. The volatile products of this action also help to maintain a protective atmospheric cover over the melt.

Sometimes an inert gas such as nitrogen or argon is piped from a tank through a graphite tube to the bottom of the crucible. This gas is allowed to bubble slowly through the molten metal, flushing the melt of dissolved gases.

An efficient but rather expensive way of obtaining gasfree castings for research is by melting and pouring metal in a vacuum. Vacuum degassing can be accomplished by melting the metal in air, placing it in an evacuating chamber, and degassing before pouring. Figure 7–16 illustrates several methods of degassing metal.

Aluminum and magnesium alloys unfortunately absorb gases quite readily. Great care must be taken to control melting, pouring, and mold conditions, and to use nothing but dry tools, ladles, furnace linings, and degassing materials. The importance of these precautions to the production of sound castings cannot be stressed too highly.

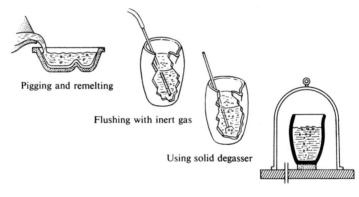

Pigging and remelting

Flushing with inert gas

Using solid degasser

Vacuum degassing

Fig. 7–16. Some methods of degassing metal.

BIBLIOGRAPHY

1. C. W. Briggs, "Solidification of Steel Castings," *Trans. AFS*, **68,** page 157, 1960
2. H. F. Bishop and W. S. Pellini, "Solidification of Metals," *Foundry*, **80,** February 1952, page 86
3. M. B. Bever, *Iron Age*, **161,** April 22, 1948
4. R. W. Ruddle, *The Solidification of Castings*, Institute of Metals, London, 1957

5. N. Chvorinov, "Theory of the Solidification of Castings," *Giesseri*, **27,** page 177–225, 1940; British Iron and Steel Institute, translation no. 117

6. H. F. Taylor, M. C. Flemings, and J. Wulff, *Foundry Engineering*, John Wiley, New York, 1959

CHAPTER 8

DESIGN OF RISERS AND
FEEDING OF CASTINGS

The primary function of a riser is to feed molten metal to the casting as it solidifies, to prevent internal or external shrinkage in the casting. Thus we may define a riser as a reservoir of molten metal which is designed to solidify last, and which is attached directly to the mass of metal of the casting. The solidification shrinkage of the casting is held to a minimum on account of the fact that the casting is fed by metal from this reservoir. If the riser is too small, shrinkage cavities occur in the casting. If the riser is too large, too much metal is melted per casting, and thus the cost of the casting is increased. The function of the riser is to supply liquid metal to the mold cavity during cooling and solidification; thus the liquid-contraction and solidification-contraction phases are materially affected by the riser. It also follows that, to ensure directional solidification, there must be proper temperature gradients.

The fact that contemporary civilization requires an infinite variety of castings has challenged metallurgists to use all their skills to develop a scientific approach to quantitative risering. The first step is to take any casting, no matter how complex, separate it into a combination of shapes, and treat each shape as a unit which solidifies as an individual casting. If one knows the solidification characteristics of simple shapes, then one need only determine which of these shapes are involved, and then design for each a riser of definite properties to feed it.

The casting must be broken down into thin and thick castings. The thin sections, of course, solidify rapidly enough; it is only the thick sections that require individual risers. One needs first to know the shrinkage characteristics of a given metal during solidification, then one must take into consideration the rate of evolution of heat by the metal as it solidifies, and the rate of heat transfer across the sand–metal interface. The problem, in essence, is to determine the smallest riser that will deliver the necessary volume of molten metal to the casting as it solidifies, and to keep this metal liquid until the casting or the section to be fed, has completely solidified.

THEORETICAL ANALYSES OF RISERING

After many studies of solidification, N. W. Chvorinov developed the logarithmic curve shown in Fig. 8–1, relating the volume and surface area of a casting to the time required for solidification. This is stated as *size coefficient R*:

$$R = \frac{\text{Volume, cu. in.}}{\text{Surface area, sq. in.}}.$$

Because a riser can be treated as a casting differing only in shape from the casting to be fed, then the solidification rate of the casting can be taken as being dependent on the relation between the surface area and the volume of the casting and riser. In order to ensure adequate feeding, the riser should freeze at a slower rate than the casting. Therefore the relative volume can be expressed as follows:

$$\text{Volume} = \frac{\text{Volume of riser as poured}}{\text{Volume of casting}}. \tag{1}$$

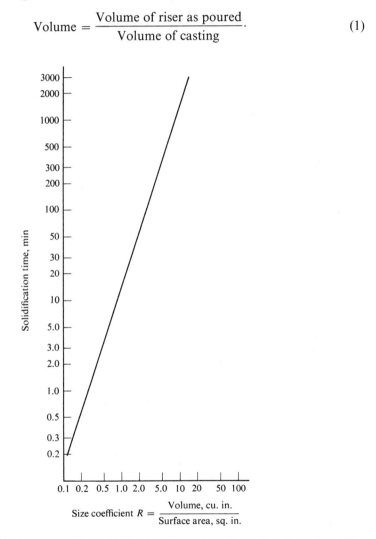

Fig. 8–1. Logarithmic curve giving solidification time of castings of various sizes (after Chvorinov).

Figure 8–2 then shows:

$$\text{Relative time needed to complete solidification (freezing ratio)} = \frac{\dfrac{\text{Surface area of castings}}{\text{Volume of castings}}}{\dfrac{\text{Surface area of riser}}{\text{Volume of riser}}}. \tag{2}$$

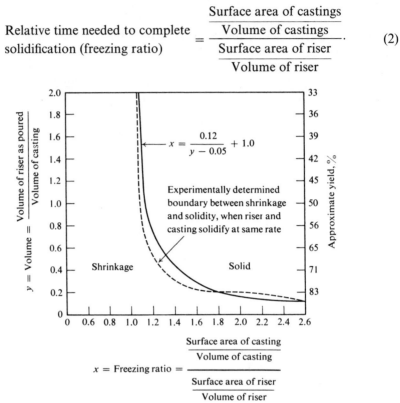

Fig. 8–2. Curves showing relationship between freezing time and volume in riser systems for steel castings.

Risering, then, is divided into two phases: (1) positioning the risers, and (2) dimensioning the risers so that they will efficiently feed the casting in general, or feed that part of the casting which they are intended to feed.

One equation for dimensioning risers is

$$x = \frac{a}{y - b} + c, \tag{3}$$

where x = freezing ratio, y = volume, and a and b are constants which relate to the mode of solidification and the contraction of the metal as it cools from the pouring temperature through the solidification range. Constant c is a measure of the difference in relative freezing rates of the riser and the casting, due to independent variables. For example, in its simplest form, for steel, it becomes:

$$x = \frac{0.12}{y - 0.05} + 1.0, \tag{4}$$

where y = volume of steel expressed as a fraction, resulting from the solution of Eq. (1), and x = relative time in units, from Eq. (2). Constant a has been assigned the value 0.12 empirically, from experiment. Constant b is the total volume of the liquid; this value varies according to the pouring temperature. Constant c is the measure of any change in the relative freezing rate of the casting and riser. If both casting and riser are in contact with sand, and are dissipating heat into the sand at the same rate, constant c is 1.0.

Solving Eq. (4) for steel results in a curve represented by the solid line in Fig. 8–2. An increase in the size of the riser means that both the x (relative freezing time) and y (volume) values increase. Those risers corresponding to points above and to the right of the curve should have sufficient volume and should freeze at a slow enough rate to ensure solidity of the casting or section they are to feed. Those risers corresponding to points below and to the left of the line would probably feed improperly, due either to insufficient volume or to too-rapid solidification.

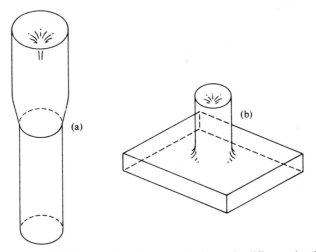

Fig. 8–3. A cylinder and plate of equal volume require risers of a different size due to the fact that the surface area of the plate is greater than that of the cylinder.

When we compare two shapes which have the same volume, we can see that the surface–area–volume ratio of the casting affects these same ratios of the riser. A cylinder such as (a) in Fig. 8–3 requires a greater riser volume than plate (b) because of the greater area of the plate, which improves the rate of heat transfer. Due to the fact that the cylinder has a concentrated mass, with less surface area to encourage a fast rate of heat transfer, the cylinder remains liquid longer. In his original study, Caine developed the equation for the freezing time of risers, as shown in Fig. 8–2. J. F. Wallace, in a further study, provided additional information by adding mildly exothermic material to the top surface of a riser and using exothermic sleeves together with the topping. This is illustrated in Fig. 8–4.

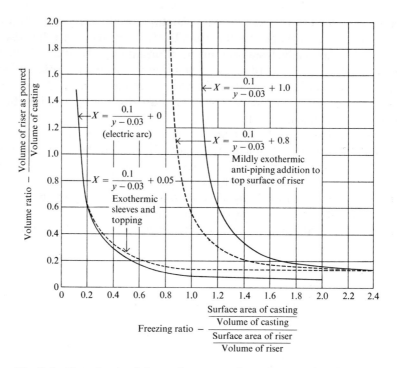

Fig. 8–4. Plot of ratio of riser volume to casting volume vs. freezing ratio,

$$\frac{\text{casting area/casting volume}}{\text{riser area/riser volume}}$$

for various conditions existing in the riser. (From J. F. Wallace, Ref. 6.)

QUANTITATIVE ANALYSES OF RISERING

Let us consider a quantitative study of riser requirements, as illustrated by the solidification of a simple casting (Fig. 8–5). Various steps of solidification of aluminum are illustrated, from completely liquid to completely solid. As solid metal is formed, contraction is evident at the top of the casting. This volumetric contraction continues as the metal's temperature goes down to 1218°F; the metal remains at this temperature until all the metal is solid. The contraction amounts to about 7% of the volume. There are solid walls and bottom, and the beginning of a solid surface on the top of the casting (Fig. 8–5d). This solid top surface, if it becomes thick enough, supports itself. Any further solidification contraction of the internal liquid metal resolves itself in an internal shrinkage cavity. The casting is completely solid, and yet is still at the temperature of 1218°F. The physical dimensions of the width are still unchanged; only the height is visibly changed. As the temperature of the casting cools down to room temperature, the casting exhibits a continued shrinkage in its width (Fig. 8–5f), a shrinkage which amounts to approximately 5% of the volume of the casting. The patternmaker takes care of this last contraction when he designs the pattern, using a patternmaker's shrinkage rule, and making the pattern 5% larger than he wishes the finished casting to be.

It is obvious that, for the first two contractions—liquid and solidification contraction—a riser is necessary. If the casting is to be sound, with no cavities present, the riser must feed approximately $7\frac{1}{2}$% of the volume of the casting. Unless the riser is considerably larger than the casting, it will freeze first. Thus, from theoretical considerations, one can establish a few guidelines by which to calculate the control of solidification.

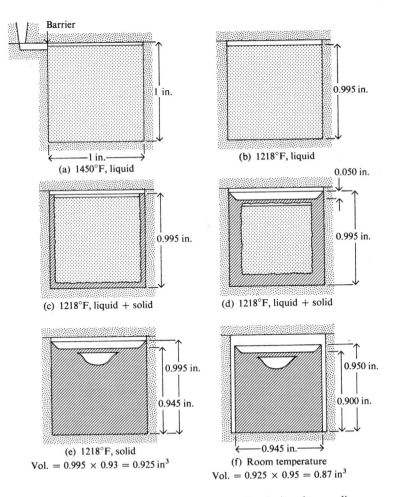

Fig. 8–5. The contraction of an aluminum cube during slow cooling.

When one is determining the rate of heat transfer, the surface area of the casting relative to its volume is important. Chvorinov expressed this mathematically as follows:

$$t = \frac{V^2}{A^2},$$

where t = the solidification time, V^2 = the square of the volume, and A^2 = the square of the area.

Since the riser should be the last to solidify, it can be considered as a casting, and if $t_r > t_c$, then it can be said that

$$\frac{V_r}{A_r} > \frac{V_c}{A_c}.$$

When we are considering the shape of the riser, and want it to have a solidification time equal to or greater than that of the casting, we realize that it should be in the shape of a sphere. Spherical shapes for a riser are not always easy to mold. Thus it is more practical to use a cylinder shape. If a blind riser (one completely enclosed in sand) is used, the top may be hemispherical in shape and thus have the smallest possible surface–area–volume ratio.

The shape of the shrinkage cavity in a riser must be watched; if there is a pipe (or an inverted conical shape) formed in the riser and it extends down into the casting, then the riser is too small, and must be enlarged.

There are various methods of calculating riser size. One is to take into consideration the shape factor of the casting, expressed as the sum of the length and width of the casting divided by the thickness:

$$\frac{L + W}{T}.$$

This approach is illustrated in part (a) of Fig. 8–6; to calculate riser diameter and height more simply and rapidly, see Fig. 8–6(b).

When a riser supplies liquid metal into the mold cavity during solidification of the casting, the contact point between the riser and the casting must not freeze before the casting does. Although a small contact point allows for easier trimming of the riser from the casting, it must be large enough to remain open while the mold is being filled and also after solidification takes place in the casting. Then directional cooling allows solidification to continue up into the riser.

Thus the foundry engineer must consider the distance through which the molten metal is fed, especially in the case of long thin sections, when feeding takes place through the end of a bar or plate. When metals are cast without a riser, the ends of the bar for a certain distance, remain sound, because there is a greater mold surface at the ends. Therefore more heat is extracted at the ends than at the center. If a long bar is cast horizontally, with a riser at the center, for a certain distance from the riser, the casting is sound. Beyond this there may be evidence of shrinkage.

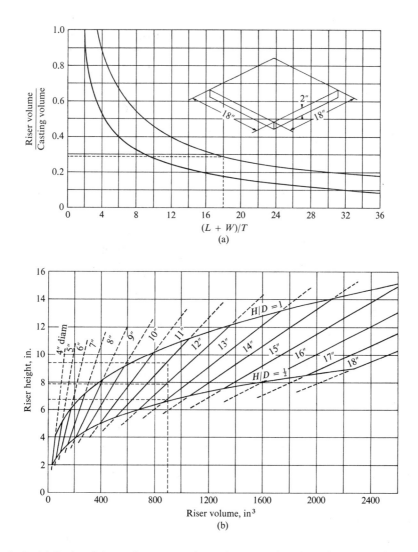

Fig. 8–6. (a) Ratio of riser-volume to casting volume as a function of the shape factor. (b) Chart for determining riser diameter. (From E. T. Myskowski, H. F. Bishop, and W. S. Pellini, Ref. 8.)

In all cases, no matter what type of shrinkage voids are formed, a casting needs an adequate reservoir of liquid metal. A reservoir such as that shown in Fig. 8–7 for area A eliminates shrinkage there. However, the same cavity will exist in B until there is a reservoir attached to B. But, with both heavy sections A and B adequately fed, then centerline shrinkage at C still remains. In order to ensure soundness at C, the casting designer should see to it that the center section tapers from the center to the heavy ends, which are adequately risered. This makes possible directional solidification from the center to each end, and up into each riser.

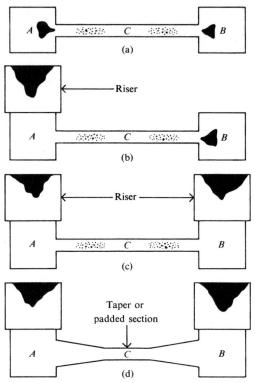

Fig. 8–7. The development (a) to (d) of a riser and padding system to ensure casting soundness. Shaded areas represent microporosity. (From J. B. Caine, Ref. 2.)

DEVELOPING THERMAL GRADIENTS

Risers are attached to heavy sections of a casting, which are the last portions of a casting to solidify. A casting which contains more than one heavy section joined by lighter sections needs a separate riser for each heavy section. If the thermal gradients within the casting are increased so that the sections farthest away from the riser are the coldest and the risers are the hottest, then the feeding distance can be increased. Several ways of increasing the thermal gradients are:

1. Using a chill
2. Padding the section of the casting nearest the riser
3. Using insulation and exothermic material around or on top of risers
4. Gating into the risers.

When the liquid metal is poured slowly, increased thermal gradients become evident. When the metal is gated directly into side risers, hot metal becomes concentrated here.

BLIND RISERS

When a casting has a minimum surface area, one can use an *enclosed* or *blind riser* (Fig. 8–8). For example, steel solidifies with a solid outer skin of metal, and the blind riser thus constitutes a closed shell of metal. The shrinkage taking place in the casting draws liquid metal from the riser, which tends to create a void, and thus a partial vacuum develops. When a core is inserted into the blind riser, atmospheric pressure becomes available to help push the metal from the riser into the shrinkage area of the casting. The core acts as a pipeline to relieve the vacuum developing because of the shrinkage of the casting and riser. Atmospheric pressure forces the liquid metal in the reservoir of the blind riser on into the casting. This method is effective only with a metal which forms a solid outer skin of metal during solidification, such as steel. If the skin of the casting is punctured, the partial vacuum is destroyed and the feeding system will not function.

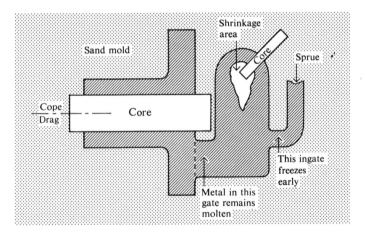

Fig. 8–8. Cross-sectional diagram of a casting fed by a blind riser with atmospheric vent produced by a pencil core.

Blind risers have several advantages: (1) They are easy to position, (2) they can be smaller than open risers, (3) they can promote directional solidification because the metal in the riser is hotter than the metal in the casting.

The size and type of connections between the riser and the casting may determine to what extent the riser can feed the casting, as well as affecting the mode of solidification and the depth of the shrinkage cavity. If the connections are allowed to solidify just after the casting solidifies and just before the riser does, this prevents

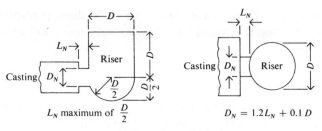

L_N maximum of $\dfrac{D}{2}$ $D_N = 1.2 L_N + 0.1 D$

(a) General type of side riser

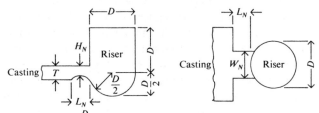

L_N maximum of $\dfrac{D}{3}$; H_N varies from 0.6 to 0.8 T ; $W_N = 2.5 L_N + 0.18D$

(b) Side riser for plate casting

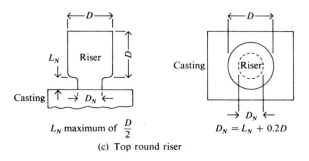

L_N maximum of $\dfrac{D}{2}$ $D_N = L_N + 0.2D$

(c) Top round riser

Fig. 8–9. Location of dimensions used in Table 8–1 for three types of risers. (From J. F. Wallace, Ref. 6.)

Table 8–1 Riser-neck dimensions*

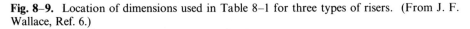

Type riser	Length L_N	Cross section L_N
General side	Short as feasible, not over $D/2$	Round, $D = 1.2 L_N + 0.1D$
Plate side	Short as feasible, not over $D/3$	Rectangular, $H_N = 0.6$ to $0.8D$; as neck length increases, $W_N = 2.5L_N + 0.18D$
Top	Short as feasible, not over $D/2$	Round, $D_N = L_N + 0.2D$

* From J. F. Wallace [6].

the shrinkage cavity in the riser from extending into the casting. Figure 8–9 gives optimum sizes for riser contacts, assuming that the same mold material is used for both riser and casting, and these data are tabulated in Table 8–1.

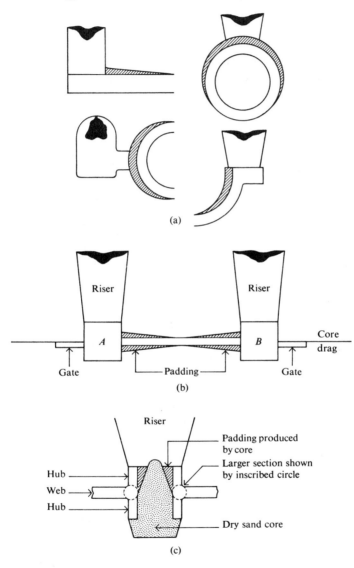

(a)

(b)

(c)

Fig. 8–10. Examples of padding. (a) Castings padded to obtain directional solidification. (b) A casting divided into two portions, each fed by individual risers. Padding has been added to help avoid shrinkage in the thinner section joining the two thicker sections. (c) Cross-sectional view through a wheel hub, in which padding has been added to help obtain progressive solidification.

PADDING

Padding is an alternative way of improving the soundness of a casting through directional solidification. A casting solidifies at its thinnest section first. If the sections of the casting are made gradually thicker toward the heavy section and the riser, the freezing progresses directionally toward it (Fig. 8–10). This padding, which is really excess metal added to the casting to develop a temperature gradient for directional solidification, can be removed by machining. However, padding often adds to the cost of the casting, so this factor must be taken into consideration when one is designing a padded casting.

EXTERNAL AND INTERNAL CHILLS

Directional solidification can also be accomplished through the use of external and internal chills. The metal at that portion of the casting which is farthest from the gating area is chilled quickly, by the use of devices called *chills*, which may be simple square or round shapes, or which may be shaped to conform to the casting dimensions. External chills are usually made of steel, iron, graphite, chromite, or copper. These chills are rammed up with the pattern and become part of the mold wall at locations selected to increase the freezing rate of the casting at this point. Figure 8–11 shows examples of external chills, which not only promote directional solidification, but engender temperature gradients in steel which reduce the possibility of microporosity.

The use of chills requires extra care, in that they must be clean and free from rust or other surface matter, and they must be thoroughly dried before they are used in a mold. If they are allowed to stand too long in the mold, condensation can occur. This produces moisture which causes gas and blowholes in the casting. If the chill is preheated before being placed in the mold or before the mold is poured, this improves conditions considerably. Figure 8–12 shows other molding materials which can be used to promote directional solidification.

Internal chills, as the name implies, are placed internally in the mold in places that are difficult to reach with external chills. They are often placed in areas in the casting that are to be machined. Frequently internal chills do not fuse completely with the casting, which naturally causes weakness at this point. The mechanical properties of the casting may be changed on account of the use of chills. Perhaps the single most important point is that the composition of internal chills must be compatible with that of the parent metal being poured. Figure 8–13 shows some examples of internal chills.

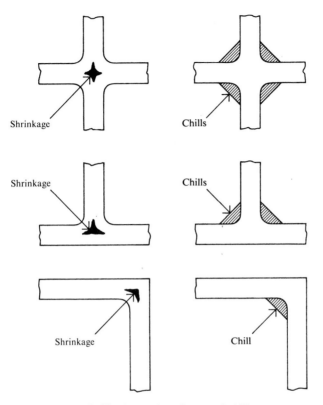

Fig. 8–11. Examples of external chills.

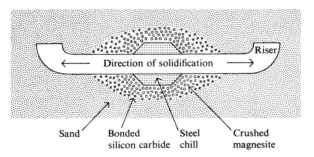

Fig. 8–12. Directional solidification can be secured by the use of a variety of mold materials that change the cooling characteristics of the mold. (From C. W. Briggs, Ref. 5.)

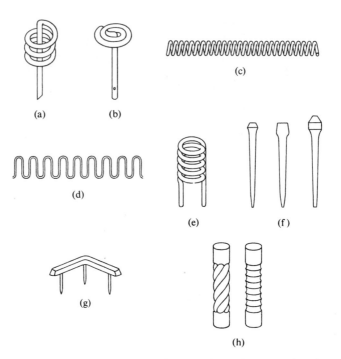

Fig. 8–13. Examples of internal chills. (a) Chill coil nail, (b) flathead chill coil nails, (c) chill coil, (d) grid chill, (e) hub chill, (f) chill nails, (g) spider chill, (h) chill rods. (Courtesy of Fanner Manufacturing Co.)

INSULATING COMPOUNDS AND EXOTHERMIC MIXTURES

The use of insulating compounds or exothermic mixtures contributes much to the effectiveness of risers. When insulating sleeves are used to form the riser wall, heat is retained in the riser for a longer period of time, allowing the riser to feed the shrinkage in the casting longer (Fig. 8–14). If ground-up insulating material or insulating substances such as rice hulls are thrown over the top of the open riser in addition to the sleeves, an even greater advantage accrues. In order to promote directional solidification, insulating pads may be cut to shape and used to form side risers or to slow the cooling rate of a thin section in a casting.

When molten metal comes into contact with an exothermic material, there is a chemical reaction which causes the exothermic material to generate heat. In some instances, the exothermic material raises the temperature of the metal at the area of contact. However, the purpose of using exothermic material is to slow down the transfer of heat from the riser to the casting. Exothermic material produces heat for only a short period of time after the metal has been poured. After that it serves as an insulator, reducing the rate of heat transfer.

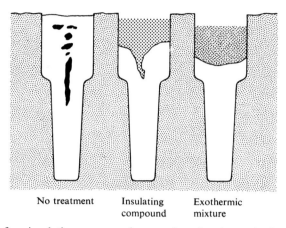

No treatment   Insulating   Exothermic
                 compound   mixture

Fig. 8–14. Use of an insulating compound or exothermic mixture in the riser reduces the piping tendency and decreases the amount of metal required in the riser.

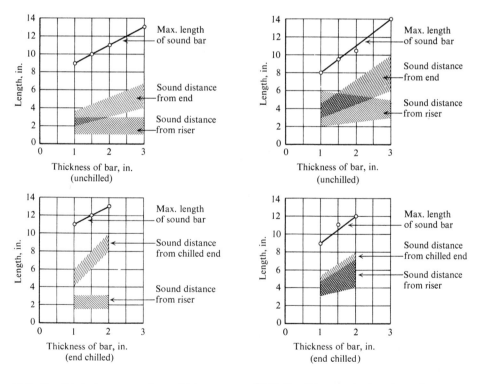

(a) Feeding distance in manganese bronze bars   (b) Feeding distance in aluminum bronze bars

Fig. 8–15. Feeding distances in manganese bronze and in aluminum bronze. (a) Feeding distance in manganese bronze bars. (b) Feeding distance in aluminum bronze bars.

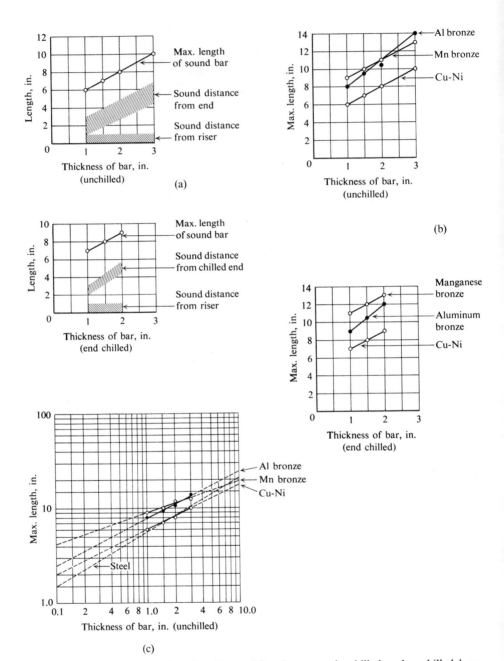

Fig. 8–16. Graphs illustrating the effects of freezing range in chilled and unchilled bars. (a) Feeding distance in 30 Ni cupronickel bars. (b) Feeding distance in bars of manganese bronze, aluminum bronze, and cupronickel. (c) Log-log plot of feeding distance for various alloys.

WIDE-RANGE VERSUS NARROW-RANGE FREEZING

In alloys which have a wide freezing range, a fine shrinkage pattern develops. By contrast, in alloys with a narrow freezing range (such as aluminum bronze and manganese bronze), shrinkage is concentrated in large voids. When the sections of the casting are made thicker, the distance of soundness of the casting (that is, the length of the casting that is sound) is greater in the aluminum bronze than in the manganese bronze. Figure 8–15(b) indicates an increase of nearly 4 inches (of unchilled bar) over that of manganese bronze (Fig. 8–15a). This effect is also found in end-chilled castings, illustrated in the same two graphs (shaded areas).

In the language of metal-casting, the word *sound* is used in a special way. A casting is considered to be sound if it is without gas holes, blow holes, cracks, segregations, and so forth. In other words, it is a satisfactory, salable casting. (This does not mean that it is radiographically perfect; that is another qualification that is sometimes applied.) In Fig. 8–15, *feeding distances* are those lengths of a casting in which feeding takes place to produce a sound distance (a sound length of casting).

Data for both plates and bars, as developed by H. A. Schwartz [Reference 8], can be represented by the equation

$$D = 5.61 T^{0.53},$$

where D = feeding distance and T = bar or plate thickness.

Figure 8–16 is a log–log plot of the data derived from the investigation, plus the data for steel. The equations for the copper alloy curves are:

Manganese bronze	$D = 9 T^{0.342}$
Aluminum bronze	$D = 5.8 T^{0.473}$
Cupronickel	$D = 7.7 T^{0.487}$

In cast iron, a self-feeding situation arises at the time the eutectic metal solidifies, because as it solidifies it expands, which partially compensates for its solidification contraction. In the case of gray iron, feed channels to the riser need only to pass liquid metal during the initial stages of solidification. As long as risers of an adequate size are used, the feeding distances in uniform sections of gray iron can be almost unlimited.

MOLD-WALL MOVEMENT WITH GRAY IRON

The movement of mold walls in gray iron castings, which produces a large effect on the final dimensions of a casting, is influenced by the high pouring temperature and the mushy mode of solidification of this metal. Due to the high pouring temperature, there is considerable thermal expansion in the sand mold. The mushy solidification mechanism results in the gray iron following the mold wall during solidification.

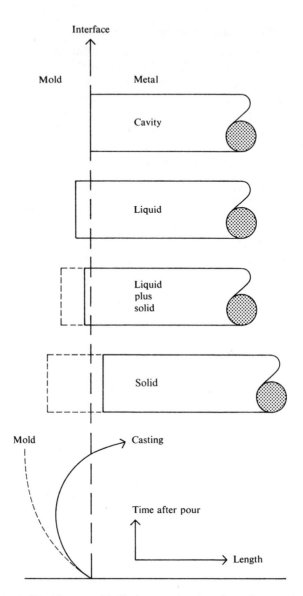

Fig. 8–17. Linear mold dilation as related to time after pouring.

Figure 8–17 diagrams the expansion of a mold for a bar of gray iron with a riser of sufficient volume to compensate for only liquid and solidification contraction. The shrinkage cavity resulting from movement of the mold wall produces an unsound casting. When there is considerable dilation of the mold wall, the riser drains, which produces a casting that is dimensionally oversized and which may have a shrinkage cavity within.

A green-sand mold undergoes an initial expansion as molten metal enters it, due to the fact that the heated inside surface of the mold expands into the high-moisture layer of sand immediately back of this layer. When the sand layer is thick enough and strong enough to resist the outward expansion, it stops. Since a dried or baked sand mold contains no moisture, there is no initial outward expansion.

Due to this sand expansion the mold cavity in a green-sand mold is larger than that in a dry-sand mold. During solidification of a cylinder $3\frac{3}{4}$ inches in diameter and 8 inches long, volumetric expansions of about $2\frac{1}{2}\%$ have been reported in green sand and compared with only about three-fourths of 1% in dry-sand molds. Under poor conditions, this expansion can be as high as 20%.

Thus we see that, to reduce the movement of mold walls in green sand, we must (1) reduce the moisture or clay content of the sand, (2) lower the temperature of the sand mold surface, (3) increase the solidification rate of the iron, (4) reduce the expansion possibilities of the sand, or (5) create a denser mold. Or a combination of these measures.

The mechanism of solidification of gray iron depends on several factors, among which are composition, cooling rate, inoculants, and melting variables. Primary austenitic dendrites contract when they solidify, but the eutectic cells, because of the low density of the graphite, expand when they solidify.

Graphitic carbon has a mushy-zone type of solidification and a pronounced tendency to follow the mold wall. The great force of expansion of a eutectic against the mold wall increases the movement of the wall. Although this expansion does influence the overall dimensions of a casting, the shrinkage cavity does not increase, nor are more risers required because of the expansion effect of solidifying graphite. A good bit of graphitic expansion goes to compensate for the volumetric expansion produced by the movement of the mold wall. The shrinkage in gray iron castings can be directly correlated with the movement of the mold wall. Expansion of the mold wall could cause shrinkage even in risered gray-iron castings, since the risers may not compensate for expansion in all sections.

BIBLIOGRAPHY

1. J. B. Caine, "A Theoretical Approach to the Problem of Dimensioning Risers," *Trans. AFS*, **56**, page 492, 1948

2. J. B. Caine, "Risering Castings," *Trans. AFS*, **57**, page 66, 1949

3. H. F. Bishop and W. S. Pellini, "The Contribution of Riser and Chill-Edge Effects to Soundness of Cast Steel Plates," *Trans. AFS*, **58**, page 185, 1950

4. W. S. Pellini, "Factors which Determine Riser Adequacy and Feeding Range," *Trans. AFS*, **61**, page 61, 1953

5. C. W. Briggs, "Risering of Commercial Steel Castings," *Trans. AFS*, **63,** page 287, 1955

6. J. F. Wallace and E. B. Evans, "Risering of Gray Iron Castings," *Trans. AFS*, **66**, page 49, 1958

7. D. Merchant, "Dimensioning of Sand Casting Risers," *Trans. AFS*, **67**, page 93, 1959

8. E. T. Myskowski, H. F. Bishop, and W. S. Pellini, "Application of Chills to Increasing the Feeding Range of Risers," *Trans. AFS*, **60**, page 389, 1952

9. R. W. Heine, C. R. Loper, Jr., and P. C. Rosenthal, *Principles of Metal Casting*, second edition, McGraw-Hill, New York, 1967

10. H. F. Taylor, M. C. Flemings, and J. Wulff, *Foundry Engineering*, John Wiley, New York, 1959

CHAPTER 9

GATING DESIGN

There are many factors that must be controlled if a good casting is to be obtained. For example, the soundness of castings can be affected by the way the molten metal enters a mold and solidifies. In order to determine the correct design of a gating system, one needs to know the flow and solidification characteristics of molten metal.

Molten metal is introduced into the mold cavity through a *gating system*, composed of four main parts: a basin, a sprue, a runner, and gates. The metal is poured through the basin into the vertical sprue, flows through a channel cut in the molding sand—called a *runner system*—and then through properly located gates into the mold cavity.

Metals in their liquid state absorb gases. They erode the mold materials during the flow process and, as they go through the process of solidification, they shrink in volume.

An ideal gating system for the production of sound metal castings must meet the following criteria.

1. The metal should flow through the gating system with a minimum of turbulence in order to avoid the oxidation of metal, the entrainment of air, the aspiration of mold gases, and the inclusion of undesired matter because of erosion or slag entrapment.

2. The metal should enter the mold cavity in a manner that will produce temperature gradients in the metal as well as on the mold surfaces, so that solidification will take place progressively in the direction of the risers or feed heads.

3. The gating system must be large enough to accommodate the casting to be poured, and yet not excessive in size, so that the desired percentage of good castings is obtained.

The first requirement must be met in order to minimize casting defects caused by inclusion of slag, dross, molding sand, and entrapped gases. The second requirement must be met in order to avoid shrinkage and defects resulting from inadequate feeding. The third requirement is a matter of economics: The casting must be produced at a cost that is competitive with other methods of manufacture.

The system should begin with the use of a pouring cup or basin (Fig. 9–1), to make it easier for the pourer to maintain a full system and provide the required flow of liquid metal. The sprue should be tapered, rather than straight, with the small end at the bottom. This minimizes vortexing of the metal at the sprue entrance.

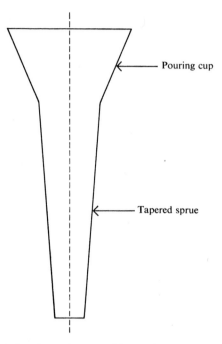

Fig. 9–1. Tapered sprue with pouring cup attached.

A basin is used to help separate the dross and slag from the metal before it flows through the system. Figure 9–2 shows the design of the pouring basin, and illustrates the flow pattern of the metal. The metal must be poured into the basin at a point that is remote from the sprue hole. If metal is poured directly down the sprue hole, vortexing and turbulence develop, creating an unsound or defective casting. But when the metal is poured into the low side of the basin, a dam effect enables the operator to reach an optimum pouring speed *before* any metal enters the sprue. This effect must be maintained to prevent dross and slag from entering the runner system. When alloys that develop large amounts of dross are poured rapidly, a bar core is sometimes used in the basin to aid the skimming action, as in Fig. 9–3.

Pouring cups, whether external or cut in the molding sand, serve only to make it easier for the operator to direct the flow of metal. They do not fulfill the functions of a good pouring basin.

A properly designed sprue is an extremely important feature in a good gating system. As the stream of molten metal moves down in the sprue, it increases in speed and becomes smaller in cross section. When a sprue has the same cross section from top to bottom, air bubbles are drawn into the stream of molten metal. This trapping of air bubbles during the pouring of metal—known as aspiration of air—can be prevented by using a tapered sprue with the small end at the bottom. Figure 9–4(a) illustrates this tapered sprue.

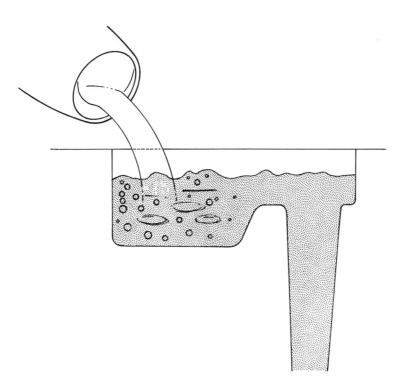

Fig. 9–2. Design of pouring basin and sprue, illustrated flow of molten metal.

Recent research on nonferrous gating indicates that good results can be obtained with rectangular tapered sprues, with a pouring cup or basin cut into the cope flat on one side near the flask and cone shaped following around (Fig. 9–4b).

Some air is almost always carried down with the first metal that enters the mold. This should be washed out in a well below the sprue. The diameter of the well should be $2\frac{1}{2}$ times the width of the runner, and the well should extend down below the runner to a depth equal to the depth of the runner. The well should have straight sides with no sharp corners, and the bottom should be as flat as possible, as illustrated in Fig. 9–5. Such a well tends to reduce turbulence and the tendency to aspirate air. It also aids in the cleanup of air entrained with the first metal that is poured.

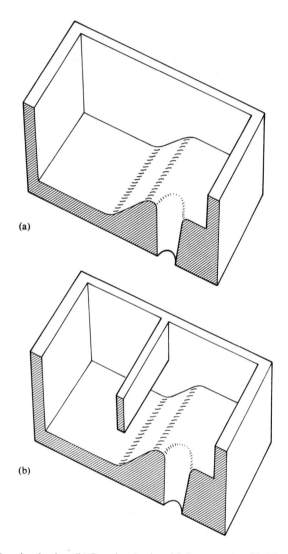

(a)

(b)

Fig. 9–3. (a) Pouring basin. (b) Pouring basin with bar core to aid skimming action.

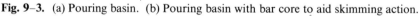

Turbulence occurs at sharp corners due to sudden changes of direction of the flowing metal, as well as sudden enlargement or contraction of the gating system. Low-pressure regions always exist at these points, and if the velocity of the metal is great enough, mold gases, from moist atmosphere, are aspirated, or drawn through the permeable mold material into the flowing stream of metal. Thus the runner system should be streamlined (Fig. 9–6), with sharp corners minimized by the use of an adequate radius. Contractions or enlargements of a runner system should be gradual or streamlined (Fig. 9–7) to prevent turbulence and low-pressure areas.

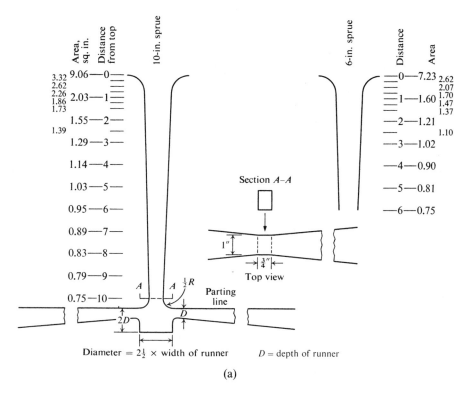

Diameter = $2\frac{1}{2}$ × width of runner D = depth of runner

(a)

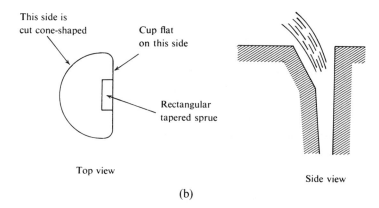

This side is cut cone-shaped

Cup flat on this side

Rectangular tapered sprue

Top view

Side view

(b)

Fig. 9–4. (a) Ideal sprue and base design. (b) Pouring basin cut into cope.

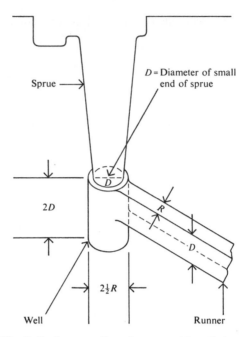

Sprue

D = Diameter of small
end of sprue

2D

R

D

Well

2½R

Runner

Fig. 9–5. Sprue, well, and runner with well sizes.

The part of a gating system that has the smallest cross section and that determines the rate of flow in the system is called a *choke* (Fig. 9–8). As the flowing metal passes through the choke area, the velocity of the molten metal increases. Increased velocity increases the turbulence in the stream of metal. A choke is used with a straight sprue to make it possible to fill the sprue quickly and keep it filled during pouring, to help lower the velocity in the runner, to float dross and slag to the cope surface of the runner, and to minimize sand erosion in the runner.

As molten metal leaves the sprue, it is traveling at its greatest velocity, and has acquired kinetic energy, which is the energy of motion. An object in motion, unless it meets with resistance, tends to remain in motion in the same direction and at the same velocity. Thus, when molten metal is flowing through a runner system, it travels at a high rate of speed, and if there is an abrupt change in direction of flow, severe turbulence may develop in the stream.

Gating systems may be grouped into two general classifications: *pressurized* and *unpressurized*. A pressurized gating system maintains a back pressure throughout, with gate and runner areas equal to or less than the area of the sprue base. For example, 1:0.75:0.5 is the ratio of sprue base area to runner area to gate area. In an unpressurized gating system, the primary restriction to the flow of the fluid is usually at the base of the sprue. This system, with a sprue base area of 1, a runner area of 2, and a total ingate area of 4 would be expressed as 1:2:4. In systems with multiple ingates, it has been found that most of the metal flows through the gates farthest from the sprue. This results from the fact that the gates and runners are incorrectly proportioned. Because of frictional losses and the abrupt changes in cross section at these points, the liquid metal has a low velocity and a fairly high pressure. Metal under these conditions flows to the farthest ingate. Less metal flows through the gates nearest the sprue because the metal has higher velocity and lower pressures there.

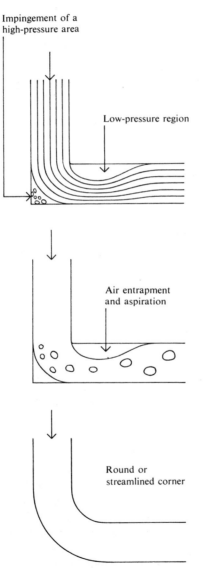

Fig. 9–6. Streamlined runner system.

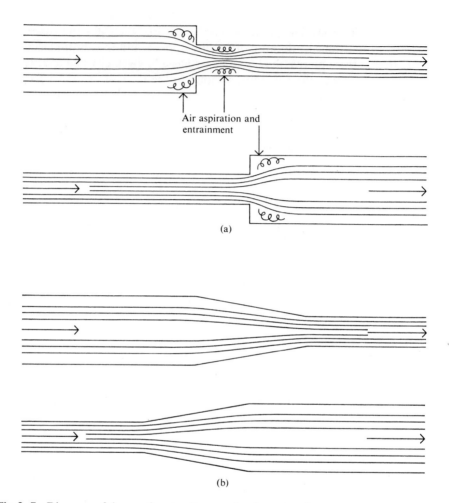

Air aspiration and
entrainment

(a)

(b)

Fig. 9–7. Diagrams of shapes of contracting or enlarging parts of a runner system. (a) Runner system which produces turbulence and aspiration of air. (b) Better design.

This is illustrated in Fig. 9–9, which shows the proportion of liquid metal flowing through the various ingates to a block mold. In this instance the ratio of total sprue area to total runner area to total gate area was 1:2:4. When a 1:2:2 gating ratio was used, an improvement was obtained, as in Fig. 9–10. But to be completely satisfactory the runner beyond each gate should be reduced in cross section, to balance the flow in all parts of the system, and thereby to further equalize the velocity and pressure. Constant pressure and velocity are maintained, while sudden changes in cross section are streamlined to reduce turbulence. Figure 9–11 illustrates such a gating system. Figure 9–12 illustrates a simple gating system for a rectangular block.

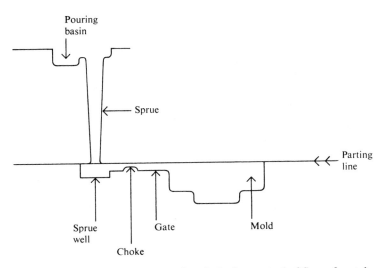

Fig. 9–8. Gating system, illustrating choke for control of flow of metal.

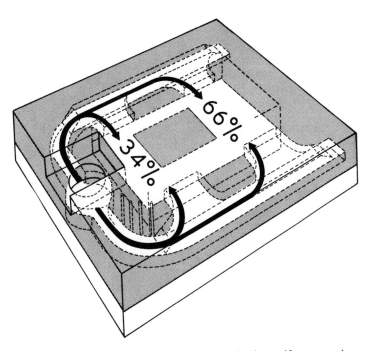

Fig. 9–9. Uneven distribution of flow in a gating system having uniform gate sizes and a ratio of 1:2:4 for total sprue area to total runner area to total gate area. (From K. Grube and L. W. Eastwood, Ref. 4.)

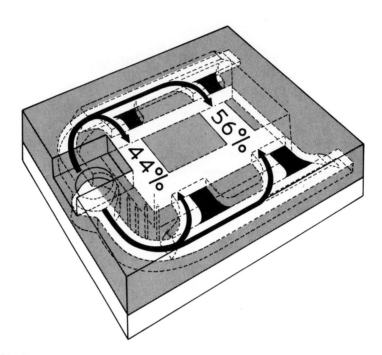

Fig. 9–10. Improved flow conditions obtained by changing sprue-to-runner-to-gate ratio 1:2:2. (From K. Grube and L. W. Eastwood, Ref. 2.)

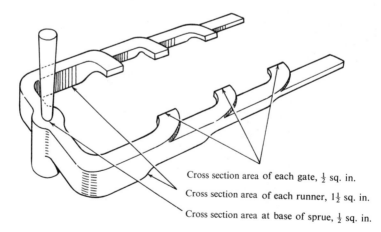

Cross section area of each gate, $\frac{1}{2}$ sq. in.

Cross section area of each runner, $1\frac{1}{2}$ sq. in.

Cross section area at base of sprue, $\frac{1}{2}$ sq. in.

Fig. 9–11. Gating system designed to give equal flow from all the gates. Gating ratio is 1:4:4 or 1:6:6 to reduce runner velocities. Cope gates make it possible for the runner to be completely filled before the metal begins to flow through the gates.

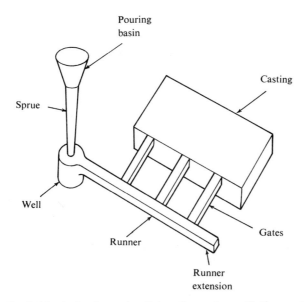

Pouring basin

Casting

Sprue

Well

Gates

Runner

Runner extension

Fig. 9–12. A simple parting-line gating system with three gates.

To increase production and improve the quality of castings produced with the use of molding machines, gating systems are prefabricated and used directly with the patterns. Or a gating system may be an integral part of a match plate, thus eliminating variations in the system. This method is far more reliable than that of hand-cutting the gating system. Once the desired runner system is determined, the designer can create a uniform system for all castings to be produced. Such a system must be carefully engineered to develop the desired characteristics. Various systems of gating are constantly being applied in the production of quality castings of all sizes.

An ideal gating system should fulfill the following functions.

1. Fill the mold cavity
2. Introduce the metal into the mold cavity with as little turbulence as possible
3. Develop the best temperature gradients in the casting
4. Control the rate of entry of metal into the mold cavity

The gating system will function properly in the production of good castings if the following aspects are carefully controlled.

1. Rate of pour into the basin (the goal being to develop a full system rapidly, and maintain it throughout the pour)
2. Size and type of sprue and runners leading to the casting
3. Type of pouring equipment, such as ladles, runner cups, and basins

4. The temperature of the metal to be poured (it must have a certain fluidity)
5. The use of risers when heavy sections are involved
6. Size and shape of gates

The leading edge of molten metal flowing in a stream follows the path of least resistance and continues to build up kinetic energy until it reaches the end of the runner. If a runner extension is used, this kinetic energy may be absorbed, thus causing a smoother flow of metal in the runners and into the mold cavity (Fig. 9–12). Sharp corners should be eliminated wherever possible in the gating system. The radius of the junction of a sprue and a runner should be carefully designed to minimize the changes in flow pattern. The entrances to mold cavities should also be well rounded to prevent the development of turbulence in the mold cavity.

Top gating is not recommended for metals which oxidize easily—such as aluminum or magnesium—since the entrapped oxides, dross, and gases resulting from turbulence produce unsound castings. This type of gating system is usually limited to small castings of simple design. However, it is also sometimes used with large billet molds which are made of metal and thus are not susceptible to erosion. The billets thus cast are rolled to produce plates and sheets. Top gating can also be used with larger ferrous castings which are not readily oxidized.

To reduce the rate of flow of metal, and to control the inclusions of slag or dross, pencil gates are often used. However, due to the characteristic dropping of the metal, severe turbulence is still present. Figure 9–13 illustrates parting-line gating and pencil gating.

The most desirable area in which to place the runner and the gates is the natural parting line of the mold. They can easily be cut or removed if they are prefabricated. It is easy to place runners in the drag and gates in the cope. Strainer screens can also easily be placed in the system and the metal can be directed into risers before it enters the mold cavity. Thus the hottest metal is in the riser, a situation which promotes directional solidification. One can control the velocity of the metal by making the cross section of the gate considerably larger than the cross section of the sprue. (We shall discuss this more fully later in this chapter.) Parting-line gates are often chosen as a molding expedient or as a compromise method of gating.

With the exception of the whirl gate, the remaining gate types (which will be illustrated in Fig. 9–19) are essentially variations of the top-gating system.

To understand the quantitative application of fluid-flow theory to gating practice, one should study turbulence, velocity calculations, and the first law of thermodynamics, or Bernoulli's theorem. Research has demonstrated that gating systems can be fully engineered by applying these fundamental principles. It is up to the foundry industry to utilize this knowledge.

Liquids flow either in a streamlined, laminar fashion or in a turbulent manner. If we treat liquid metal as a flowing liquid subject to all fluid-flow theories, we can then apply this principle to gating practice. The widespread work on gating of molten metal has shown that with liquid metal, as with other fluids, either laminar or turbulent flow is possible, depending on the velocity of flow, the viscosity, and the cross section of the flow channel. Laminar flow is streamlined, with flow occurring in well-defined, nonintersecting paths without turbulence. In turbulent flow, the paths of the liquid particles cross and recross one another in an interlacing pattern, with eddies. Very little air aspiration or drossing occurs in laminar flow. However, the velocity required for this type of flow is low; the rate of flow necessary for the gating of most molten metal usually produces some turbulence. A small amount of turbulence is allowable, since a small amount does not bring about enough metal oxidation or air aspiration to produce faulty castings. The amount of turbulence that can be tolerated varies with the reactivity of the metal and the quality desired of the casting.

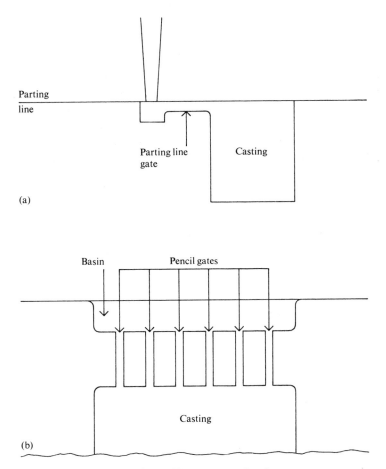

Fig. 9–13. Different types of gating.

Whether the flow is turbulent or smooth depends on the velocity of the liquid, the cross section of the flow channel, and the viscosity of the liquid. The relationship is expressed as the Reynolds number, R_n:

$$R_n = \frac{\text{velocity of flow} \times \text{diameter of channel} \times \text{density of liquid}}{\text{viscosity of the liquid}}$$

or

$$R_n = \frac{Vdp}{u},$$

where R_n = Reynolds number
  V = velocity of flow in feet per second
  d = diameter of the channel in feet
  p = fluid density in pounds per cubic feet
  u = the viscosity of the fluid in foot-seconds per pound

When the Reynolds number remains less than 2000, true streamline flow results. When the Reynolds number is more than 2000, turbulent flow prevails. Most metals reach turbulent flow conditions quite readily. Steel, which has a Reynolds number in excess of 3500, always flows under turbulent conditions. The turbulence found in well-designed gating systems does not, however, appear to be harmful to the quality of the metal, although excess turbulence creates such problems as inclusion of dross or slag, aspiration of air into the metal, erosion of the mold wall, and roughening of the casting surface. Thus gating systems are designed to reduce turbulence to a point where it is not harmful.

When you utilize hydraulic principles, you can reduce turbulence by avoiding sharp changes in cross section of flow channels and abrupt changes in direction of flow. However, all portions of the gating system should run filled with fluid, under a pressure slightly higher than atmospheric, to avoid aspiration of air into the system. Aspirated air is undesirable in molten metals because it produces increased drossing, and may result in air bubbles being entrapped under cores or on the cope surface. Figure 9–14 illustrates the effects of pressure head and change in gate design on the velocity of flow of metal.

In the design of gating systems, one uses two basic fluid-flow equations to take velocity into account. The first is the law of continuity, by which one can calculate the velocity and the rates of flow of liquid metal. This may be written

$$Q = A_1 V_1 = A_2 V_2,$$

where Q = rate of flow of liquid metal in cubic feet per second
  A_1 = cross-sectional area of flow channel at point 1 in square feet
  V_1 = velocity of liquid metal at point 1 in feet per second

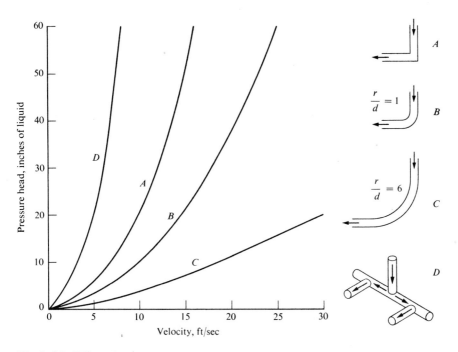

Fig. 9–14. Effect of pressure head and change in gate design on velocity of metal flow.

Since liquids are incompressible, flow rate Q must be the same in all portions of a fluid system at the same time. Figure 9–15 illustrates the use of the above equation to predict flow velocities in simple runners. When the cross section of the runner is reduced to half its original size, velocity of the metal is doubled.

The basic law of hydraulics, known as *Bernoulli's theorem*, gives relationships between the factors that influence the fluid behavior of molten metal. One can apply this theorem to gating systems with satisfactory results. The theorem states that, at any location in a system, the sum of potential head (head = height above a given reference plane), pressure or metallostatic head, and velocity or kinetic head equals a constant. To derive the equation below, we may sum up Bernoulli's theorem in terms which state that the sum of the potential energy, the velocity energy, the pressure energy, and the frictional energy of a flowing liquid is equal to a constant. When energy losses occur due to turbulence and friction, this loss must also be considered.

The factors involved in Bernoulli's theorem are shown schematically in Fig. 9–16. The potential energy is naturally a maximum at the highest point in the system, or at the top of the pouring basin. As the metal passes through the mold system, the potential energy is rapidly changed to kinetic or velocity energy and pressure energy. Once flow is established, the potential and frictional heads are virtually constant; the velocity is high when the pressure is low, and vice versa.

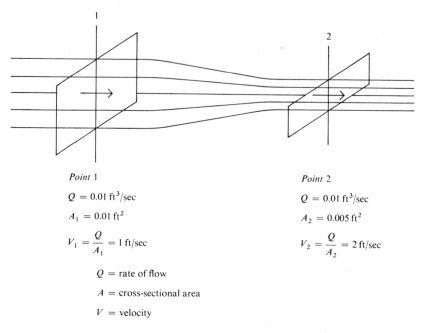

Point 1

$Q = 0.01 \text{ ft}^3/\text{sec}$

$A_1 = 0.01 \text{ ft}^2$

$V_1 = \dfrac{Q}{A_1} = 1 \text{ ft/sec}$

Point 2

$Q = 0.01 \text{ ft}^3/\text{sec}$

$A_2 = 0.005 \text{ ft}^2$

$V_2 = \dfrac{Q}{A_2} = 2 \text{ ft/sec}$

$Q = $ rate of flow

$A = $ cross-sectional area

$V = $ velocity

Continuity: $Q = A_1 V_1 = A_2 V_2$

Fig. 9–15. Predicting velocities of flow in runners.

While metal is flowing there is a constant loss of energy in the form of fluid friction between the metal and the mold wall. (Though Bernoulli's theorem does not take it into account, there is also a heat loss, which reduces the solidification time of the metal.) The theorem can be expressed as

$$wZ + wPv + \frac{wV^2}{2g} + wF = K,$$

where $w = $ total weight of fluid flowing, lb
$Z = $ height of liquid, ft
$P = $ static pressure in liquid, lb/ft^2
$v = $ specific volume of liquid, ft^3/lb
$g = $ acceleration due to gravity, 32 ft/sec/sec
$V = $ velocity, ft/sec
$F = $ frictional losses, ft
$K = $ a constant

When we divide the equation by w, all the terms have the dimensions of length, and may be considered to represent:

Potential head, Z
Pressure head, pv

Velocity head, $V^2/2g$
Frictional loss of head, F

It is most important to be able to establish the proper flow system rapidly. This means that when liquid metal enters the sprue, it should be flowing under conditions that resemble those which are present when a full flowing system has been established.

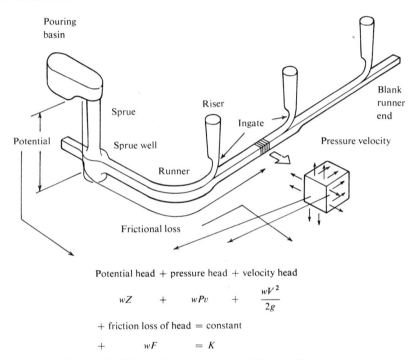

Potential head + pressure head + velocity head

$$wZ \quad + \quad wPv \quad + \quad \frac{wV^2}{2g}$$

$$+ \text{ friction loss of head} = \text{constant}$$

$$+ \quad wF \quad = K$$

Fig. 9–16. Schematic representation of Bernoulli's theorem.

Figure 9–17 shows how to predict the velocity of flow in an idealized gating system. The flow velocity is to be calculated at point B, where the ingate enters the mold cavity. We take the height of the sprue as 1 foot, and consider that there is no energy loss in the system. At point A, where the metal is exposed to the atmosphere, we consider the pressure to be 1 atm, and the velocity of flow in a large pouring basin to be zero. At point B, the pressure is also 1 atm while the metal is entering the mold, and the height is arbitrarily taken as zero. If we apply these constants to the equation, the pressure and initial velocity drop out, so that $V_B = \sqrt{2gh}$. The calculated velocity V_B is 8 ft/sec.

Let us assume that the exit area is 0.01 ft². Then

$$Q = A_B V_B = (0.01)(8),$$
$$Q = 0.08 \text{ ft}^3/\text{sec}.$$

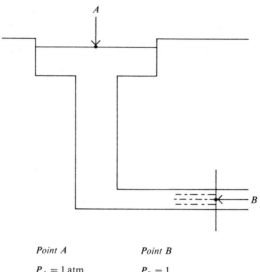

Point A   Point B

$P_A = 1$ atm $P_B = 1$

$V_A = 0$ $V_B = ?$

$h_A = 1$ ft $h_B = 0$

$$\frac{V_B^2}{2g} = h_A$$

$$V_B = \sqrt{2gh}$$

$$V_B = 8 \text{ ft/sec}$$

Fig. 9–17. Calculation of velocity of flow in an ideal sprue-runner system (no energy losses); P = pressure, V = velocity, h = metal head.

We have ignored the friction of the metal flowing in the runner passage and the internal friction (viscosity) of the metal; thus the values obtained by the above equation will be somewhat higher than values encountered in actual practice.

Occasionally you may want to cast metal in a vertical plane. Shell molds, permanent molds, and some sand molds are molds in which castings are best cast on edge, with a vertical gating system. Figure 9–18 gives examples of gating systems suitable for light metal alloys.

If you do use vertical gating, however, you need some method of dissipating the kinetic energy of the liquid flowing down the sprue, and also of reducing the velocity of the stream of metal before it leaves the gate. Certain parts of the gating system should be designed specifically to reduce the amount of turbulence in the flowing metal. For one thing, the sprue should be tapered, to prevent the entrainment of air in the stream of metal. And when the molten metal is poured down the sprue, it flows into the larger base, which reduces its velocity. From there the metal flows out into the runner. There it loses still more velocity, and any damaged part of the metal rushes on to be trapped in the runner extension. The sound metal then flows through the gates into the mold cavity, where it solidifies.

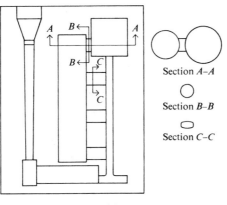

(a)

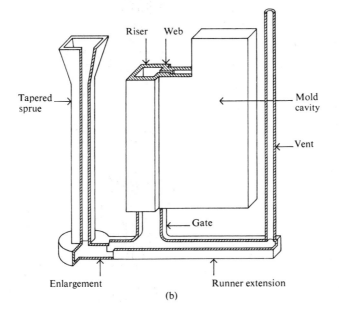

(b)

Fig. 9–18. Examples of vertical gating systems. (a) Ductile iron in shell molds. (b) Light-metal alloy.

Vertical pouring often requires the use of longer sprues which increase the exit velocity of the molten metal. If the metal enters the mold cavity directly from the

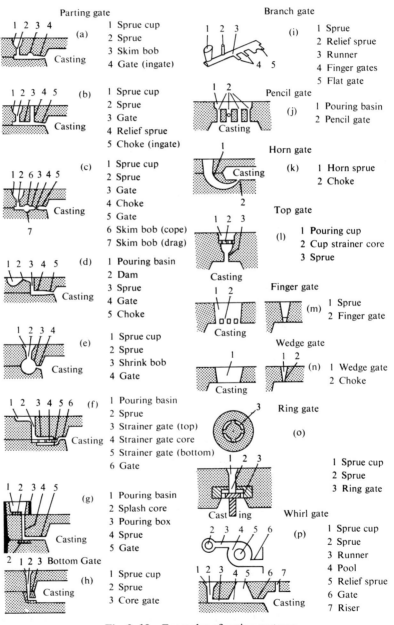

Fig. 9–19. Examples of gating systems.

sprue, a jet effect with excessive turbulence will develop. The best practice is to join the sprue to the bottom of a side riser that connects with the mold cavity throughout its entire height by means of a narrow well or slot. One must take care in designing the side riser; it should be of sufficient size to prevent metal from cascading into the mold cavity and causing excessive turbulence and entrapment of air. The sprue should discharge into a horizontal runner which, in this case, is beneath both the riser and the mold cavity.

A runner whose cross-sectional area is twice that at the bottom of the sprue reduces the velocity of the liquid and helps eliminate turbulence. If a runner has an extension past the gate, this acts as a trap for impurities, since the leading edge of the first liquid metal is likely to contain dirt. The cross-sectional area of the extension should not be greater than the cross-sectional area of the runner. A flared gate with an exit opening twice as large as the entrance opening reduces the velocity of the stream of flowing metal, and thus helps eliminate turbulence. The use of a side riser prevents turbulent liquid from entering the mold cavity. It also acts as a reservoir for the last liquid poured, and feeds the shrinkage cavity in the casting during solidification. A web connection between the side riser and the mold cavity serves to introduce the liquid quietly and progressively from the bottom to the top of the casting. If the web and riser are relatively large, this will reduce cascading and turbulence of the flowing metal.

Figure 9–19 presents fifteen examples of various kinds of gating systems.

BIBLIOGRAPHY

1. R. W. Heine and P. C. Rosenthal, *Principles of Metal Casting*, McGraw-Hill, New York, 1955

2. K. Grube and L. W. Eastwood, "A Study of the Principles of Gating," *Trans. AFS*, **58**, page 76, 1950

3. R. F. Polich, A. Saunders, Jr., and M. C. Flemings, "Gating Premium-Quality Castings," *Trans. AFS*, **71**, page 418, 1963

4. L. W. Eastwood, "Symposium on the Principles of Gating," American Foundrymen's Society, page 25, 1951

5. J. F. Wallace and E. B. Evans, "Principles of Gating," *Foundry*, **87**, page 74, October 1959

6. H. F. Taylor, M. C. Flemings, and J. Wulff, *Foundry Engineering*, John Wiley, New York, 1959

7. K. R. Grube and J. G. Kura, "Principles Applicable to Vertical Gating," *Trans. AFS*, **63**, page 35, 1955

8. K. R. Grube, R. M. Lang, and J. G. Kura, "Modifications in Vertical Gating Principles," *Trans. AFS*, **64**, page 54, 1956

9. F. E. Murphy, G. J. Jackson, and R. A. Rosenberg, "Bronze Valve Vertical Gating in Shell Molds," *Modern Castings*, **40**, page 81, July 1961

CHAPTER 10

NONFERROUS METALS: THEIR PROPERTIES AND USES

Nonferrous alloys are classified according to the base elements of which they are composed. The base elements used commercially are mainly copper, aluminum, magnesium, lead, tin, and zinc. Most commercial cast nonferrous alloys are binary, which means that they contain two principal elements with two or three additional elements in small amounts. To obtain the desired physical and mechanical properties, one must be able to manipulate the amounts of these trace elements. When a third element of more than a few percent is included, the alloy becomes a *ternary* alloy.

Nonferrous alloys are cast into many shapes and end products. Different casting processes are used, depending on the size, shape, production, and quality of castings desired.

Copper-base alloys have been classified by the American Society for Testing Materials. The classifications, with the basis for each, are given in Table 10–1.

Table 10–1 Classification of cast copper-base alloys

Class	Addition elements	Remarks
	Copper	
Copper	Not more than 2% total of arsenic, zinc, cadmium, silicon, chromium, silver, or other elements.	Conductivity copper castings, pure copper, deoxidized copper, and slightly alloyed copper.
	Brasses	
Red brass	2–8% zinc. Tin less than zinc. Lead less than 0.5%.	Alloys in this class without lead seldom used in foundry work.
Leaded red brass	2–8% zinc. Tin less than 6%, usually less than zinc. Lead more than 0.5%.	Commonly used foundry alloys. May be further modified by addition of nickel. See ASTM Specifications B62 and B145.
Semi-red brass	8–17% zinc. Tin less than 6%. Lead less than 0.5%.	Alloys in this class without lead seldom used in foundry work.

Table 10–1 (*continued*)

Class	Addition elements	Remarks
Leaded semi-red brass	8–17% zinc. Tin less than 6%. Lead more than 0·5%.	Commonly used foundry alloys. May be further modified by addition of nickel. See ASTM Specification B145.
Yellow brass	More than 17% zinc. Tin less than 6%. Less than 2% total aluminum, manganese, nickel, iron, or silicon. Lead less than 0.5%.	Commonly used foundry alloy.
Leaded yellow brass	More than 17% zinc. Tin less than 6%. Less than 2% total aluminum, manganese, nickel, or iron. Lead more than 0.5%.	Commonly used foundry alloy. See ASTM Specification B146.
High-strength yellow brass (manganese bronze)	More than 17% zinc. More than 2% total of aluminum, manganese, tin, nickel, and iron. Silicon less than 0·5%. Lead less than 0·5%. Tin less than 6%.	Commonly used foundry alloys under the name of "manganese bronze" and various trade names. See ASTM Specification B147.
Leaded high-strength yellow brass (leaded manganese bronze)	More than 17% zinc. More than 2% total of aluminum, manganese, tin, nickel, and iron. Lead more than 0.5%. Tin less than 6%.	Commonly used foundry alloys. See ASTM Specifications B132 and B147.
Silicon brass	More than 0.5% silicon. More than 5% zinc.	Commonly used foundry alloys. See ASTM Specification B198.
Tin brass	More than 6% tin. Zinc more than tin.	Alloys in this class seldom used in foundry work.
Tin-nickel brass	More than 6% tin. More than 4% nickel. Zinc more than tin.	Alloys in this class seldom used in foundry work.
Nickel brass (nickel silver)	More than 10% zinc. Nickel in amounts to give white color. Lead less than 0.5%.	Commonly used foundry alloys, sometimes called "German silver."
Leaded nickel brass (leaded nickel silver)	More than 10% zinc. Nickel in amounts sufficient to give white color. Lead less than 0.5%.	Commonly used foundry alloys, sometimes called "German silver." See ASTM Specification B149.

Table 10–1 (*continued*)

Class	Addition elements	Remarks
	Bronzes	
Tin bronze	2–20% tin. Zinc less than tin. Lead less than 0.5%.	Commonly used foundry alloys. May be further modified by addition of some nickel or phosphorus, or both. See ASTM Specifications B22 and B143.
Leaded tin bronze	Up to 20% tin. Zinc less than tin. Lead more than 0.5%, less than 6%.	Commonly used foundry alloys. May be further modified by addition of some nickel or phosphorus, or both. See ASTM Specifications B61 and B143.
High-leaded tin bronze	Up to 20% tin. Zinc less than tin. Lead more than 6%.	Commonly used foundry alloys. May be further modified by addition of some nickel or phosphorus, or both. See ASTM Specifications B22, B66, B67 and B144.
Lead bronze	Lead more than 30%. Zinc less than tin. Tin less than 10%.	Used for special bearing applications.
Nickel bronze	More than 10% nickel. Zinc less than nickel. Less than 10% tin. Lead less than 0.5%.	Commonly used foundry alloys. Sometimes called "German silver" or "nickel silver."
Leaded nickel bronze	More than 10% nickel. Zinc less than nickel. Less than 10% tin. Lead more than 0.5%.	Commonly used foundry alloys. Sometimes called "German silver" or "nickel silver." See ASTM Specification B149.
Aluminum bronze	5–15% aluminum. Up to 10% iron, with or without manganese or nickel. Less then 0.5% silicon.	Commonly used foundry alloys. Some may be heat-treated. May be further modified by addition of some nickel or tin, or both. See ASTM Specification B148.
Silicon bronze	More than 0.5% silicon. Not more than 5% zinc. Not more than 98% copper.	Commonly used foundry alloys. Some are readily heat-treated. See ASTM Specification B198.
Beryllium bronze	More than 2% beryllium or beryllium plus metals other than copper.	Most of these alloys are heat-treatable.

Alloys of copper and zinc are commonly classified as *brass*. Copper-zinc alloys are one of the most useful groups of metals, due to their desirable properties and relatively low cost. The mechanical properties of these brasses depend largely on the zinc content. The photomicrograph in Fig. 10–1 shows a treelike structure or dendritic form typical of copper-zinc alloys in the as-cast form. Brass alloys have good ductility, malleability, strength, and corrosion resistance. With increasing additions of zinc, the color of the alloy changes from copper-red to yellow at about 38 % zinc.

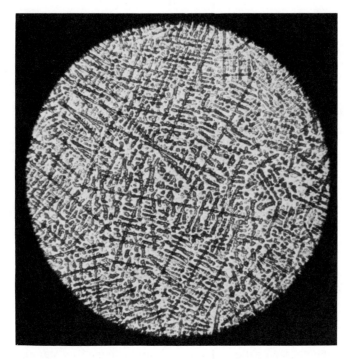

Fig. 10–1. Example of cored dendritic structure. Photomicrograph of 87-10-1-2 as-cast alloy characteristic of casting section cooled fairly rapidly, hence the fine dendritic structure. Ammoniacal copper chloride etch, 50 ×. 87 % Cu, 10 % Sn, 1 % Pb, 2 % Zn.

Yellow brass is the most ductile of all the brasses. Its ductility makes possible the use of this alloy for jobs requiring the most severe cold-forging operations, such as deep-drawing, stamping, and spinning. Figure 10–2 shows the equilibrium diagram for copper-zinc.

Lead may be added to brass to improve its cold-working characteristics. When lead is added to brass, however, it does not respond well to hot-working.

Red brass is composed of between 2 and 8 % zinc, has a reddish color, a great resistance to corrosion, and good workability. Red brass alloys have good casting

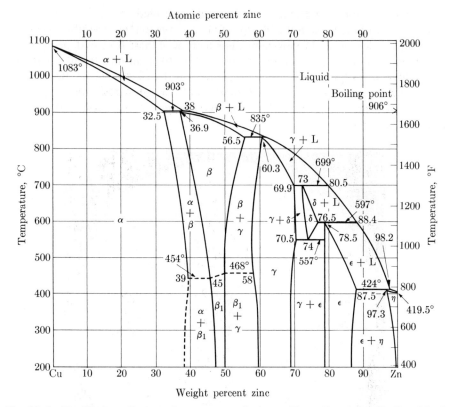

Fig. 10–2. Equilibrium diagram for copper and zinc. (From Lawrence H. Van Vlack, *Materials Science for Engineers*, Addison-Wesley, 1970.)

and machining characteristics. They are readily shaped by stamping, drawing, forging, and spinning. Applications of red brass include valves, fittings, rivets, radiator cores, plumbing pipe, flexible hose, screen cloth, and many others.

Leaded red brasses are the most popular of the cast-brass alloys (Fig. 10–3). They are used when good machinability, moderate strength, ductility, and good castability are required.

When tin is added to copper as a secondary alloying element, *bronze* is produced. Hardness and resistance to wear are increased, although ductility is decreased. The range of the tin addition is on the order of 5 to 10%. The higher the tin content, the stiffer and harder the alloy. Pure copper with 20% tin added produces bell metal, which is very hard and produces a decided ring when struck. Figure 10–4 shows the equilibrium diagram for copper and tin, and Fig. 10–5 is a photomicrograph of this alloy, which is reddish in color. Zinc up to 5%, which has

Fig. 10–3. Photomicrograph of as-cast 85-5-5-5 alloy. The white areas show a higher copper and lower tin content than the areas shown in halftone, which are just the reverse. The black spots are lead particles. Ammonia-hydrogen peroxide etch followed by ferric chloride, 100 ×.

a hardening effect, may be added to it, and lead, in amounts of 1 to 10%, may be added to improve machinability. Tin bronzes have excellent resistance to corrosion and good properties at elevated temperatures. When lead is added, an excellent bearing bronze is produced. (*Bearing bronze* is the name given to a particular alloy which, because of its strength, is quite useful as a weight-bearing material; it is frequently used to support a rotating shaft.)

Yellow bronze, with more than 17% zinc and more than 2% aluminum, manganese, iron, and tin, is known as *manganese bronze*. Manganese and iron, which are added to obtain the desired physical properties, can cause tensile strengths from 60,000 to 120,000 psi to be obtained. Lead is kept at a minimum, due to the fact that aluminum is present in the alloy. (Lead and aluminum do not enter into a solid solution, but solidify with the formation of segregations; thus, to maintain proper dispersion of the elements of the alloy, lead is generally not added to an alloy in which aluminum is an ingredient.)

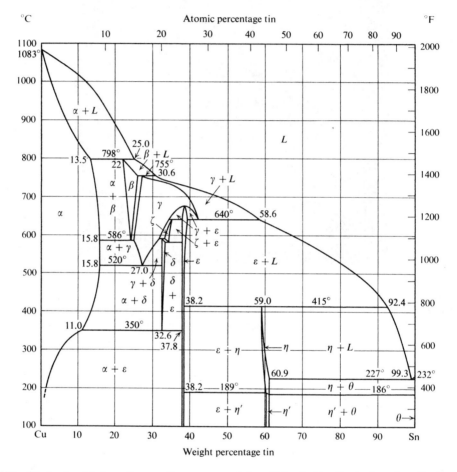

Fig. 10–4. Equilibrium diagram for copper and tin. (*Metals Handbook*, American Society for Metals.)

Bronzes with a fairly high percentage of manganese have good resistance to corrosion and great strength. They are used for propellers, lever arms, pump bodies, valves, and valve stems.

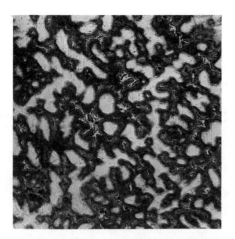

Fig. 10–5. Photomicrograph of 88-10-0-2 as-cast alloy which shows delta compound as light-colored areas in black areas of lower tin content. Ferric chloride etch, 100 × .

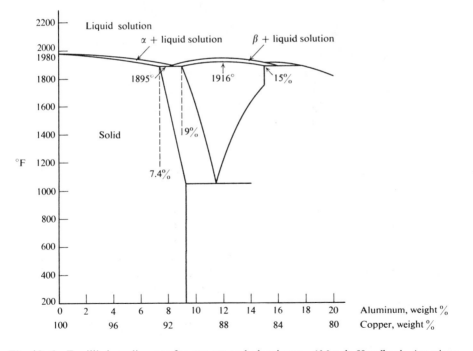

Fig. 10–6. Equilibrium diagram for copper and aluminum. (*Metals Handbook,* American Society for Metals.)

Copper-rich alloys of aluminum and copper are known as *aluminum bronze*. They contain 4 to 13 % aluminum and 1 to 4 % iron as alloying elements. Figure 10–6 shows the equilibrium diagram for these alloys. Aluminum bronze has a near-gold color, possesses a good finish, and is highly resistant to acid. A photomicrograph of it is shown in Fig. 10–7.

Fig. 10–7. Photomicrograph of as-cast aluminum bronze. Alpha matrix white, beta dark gray; iron appears in the form of rosettes. Ferric nitrate etch, 100 ×. This alloy has 10.5 % Al and 3.5 % Fe. (Courtesy Ampco Metal, Milwaukee, Wis.)

These Cu-Al alloys are more difficult to machine, however, than brass or bronze alloys. Fortunately they can be heat-treated to improve their physical properties. It is the addition of aluminum that makes these castings hard.

Adding iron also has some effect on hardness, and when castings are not subject to heat treatment, the addition of iron greatly increases the ductility of the alloy. This bronze alloy is used for gears, propellers, pump parts, and high-strength bearings that are subjected to shock.

A variation of this is a copper-aluminum alloy called *Superston*. Its normal composition is 75 % Cu, 8 % Al, 12 % Mn, 3 % Fe, and 2 % Ni. It has 94,000 to 105,000 psi tensile strength, 40,000 to 49,000 psi yield strength, and can withstand a 20 to 30 % elongation. It is used with great success in very large propellers. It has great fatigue strength in air and sea water, excellent resistance to corrosion, and very good resistance to cavitation erosion. It also has excellent foundry characteristics, and accepts hot-working and welding without becoming prey to hot short and surface cracks. (See Glossary of Terms for definition of "hot short.")

One bronze, which is red in color, contains 1 to 5% silicon, and is known as *silicon bronze*. This alloy has good resistance to salt water, alkalies, and acids. Small amounts of zinc, tin, and iron may also be present. Figure 10–8 gives the Cu-Si equilibrium diagram.

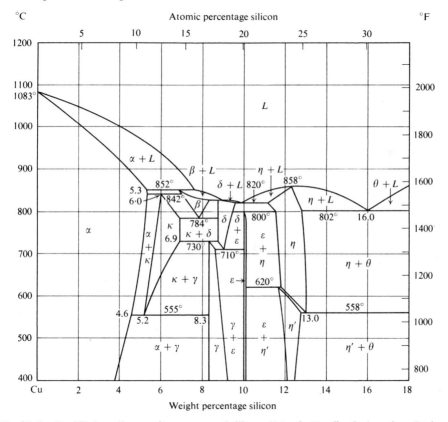

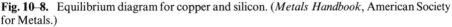

Fig. 10–8. Equilibrium diagram for copper and silicon. (*Metals Handbook*, American Society for Metals.)

Nickel-base alloys have great resistance to corrosion in the presence of most mineral acids, most organic acids, and all alkalies. They are not resistant to corrosion by nitric acid, or by oxidizing salts, such as ferric sulfates or copper sulfates. They have good mechanical strength, ductility, and resistance to wear, although they cannot be used for bearings, except under light loads and at slow speeds.

An alloy of 70% nickel and 30% copper is known as *Monel*. When silicon is added to it, it becomes age-hardened, and thus more wear-resistant. An alloy of 80% nickel and 20% chromium, known as *Nichrome*, is used for electrical resistance heaters. An alloy of 80% nickel, 14% chromium, and 6% iron, known as *Inconel*, is used where oxidation resistance with high strength at elevated temperatures is needed.

Alloys with a high percentage of nickel are used for chemical equipment, such as implements used in the dyeing of textiles and the manufacture of caustics, as well as for making water-softening equipment, valves and pump parts, and food-handling equipment.

Zinc-base alloys are widely used in die-casting. The alloying elements are principally copper, aluminum, and magnesium. The amount of each used—usually between 4 and 8%—depends on the properties desired. High-purity zinc is used as the base metal. The addition of copper increases the strength, but reduces ductility. Addition of aluminum improves the strength of the alloy and delays the rate of attack of the alloy on steel dies, and thus improves the life of the die. Additions of magnesium add to the dimensional stability of a die casting.

An alloy consisting of 20% zinc, 20% manganese, 1% aluminum, and the balance copper produces a white-bronze casting that is strong, ductile, and corrosion-resistant. It can be polished to a high, silvery luster which makes it useful in architectural and marine hardware, plumbing fixtures, ornamental castings, hospital equipment, and swimming-pool equipment.

ALUMINUM ALLOYS

Pure aluminum is a poor casting material; therefore almost all aluminum castings are really made of an alloy of aluminum. There are many aluminum alloys. Thus it follows that there are many different properties that result from the composition of the various alloys.

The selection of a particular alloy depends on the criteria demanded of the alloy: mechanical strength, machinability, surface appearance, resistance to corrosion, conductivity, and leak tightness. Tables 10–2 through 10–7 show alloys with various chemical composition and properties.

The principal alloying elements are copper, silicon, magnesium, zinc, chromium, manganese, tin, and titanium. Iron is often present in small quantities, and is considered an impurity.

The effect of the added elements is to increase fluidity, machinability, strength, and hardness of the aluminum, and to lower its ductility, conductivity, and increase its resistance to impact and resistance to corrosion.

The mechanical properties of aluminum alloys depend on alloying, heat-treatment, and the casting process. The tensile properties vary from 17,000 to 43,000 psi tensile strength, from 9000 to 43,000 yield strength, and 0 to 12% elongation capability.

Table 10–2 Uses of copper alloys

Uses	Nominal compositions, %				
	Cu	Sn	Pb	Zn	Other
Aircraft castings	Aluminum bronze Manganese bronze				
Andirons	Yellow brass				
Architectural	85 Yellow brass Nickel silver Manganese bronze	5	5	5	
Ball bearings races	Manganese bronze Aluminum bronze				
Ball retainers	87	5	2	5	0–1 Ni
Band instruments	Yellow brass Nickel silver				
Bearings	88	6.5	1.5	4	
(High speed–low load)	83	7	7	3	
Electrical machinery	85	5	5	5	
Engine	85	10	5		
Locomotive	75	5	20		0–1 Ni
Machine tool	78	7	15		
Pump and seals	72	7	20	1	
Railroad	65	5	30		
Bearings:	66		34		
(Med. speed and load)	80	10	10		
Cross head	72	7	20	1	
Locomotive	88	10	2		
Machine tool	85	5	9	1	
Main	70	5	25		
Pump	78	7	15		
Railroad	80	8	12		
Rolling mill	68	7	25		
Bearings:	Rem.	6–11	0–1	0–4	0–5 Ni
(Low speed–heavy load)					
Bridge	83	14	3		
	81	19			
	Manganese bronze				
Railroad	85	10	5		
Rolling mill	86 Aluminum bronze	10		2	1–2 Ni
Trunnion	88	10	2		
Turntable	80 Manganese bronze Aluminum bronze Silicon bronze	20			

Table 10–2 Uses of copper alloys (*continued*)

Uses	Cu	Sn	Pb	Zn	Other
				Nominal compositions, %	
Bearings:	70	5	25		
Self-lubricating	80	10	10		Graphite
Slow speed–heavy load	Manganese bronze				Graphite
	Aluminum bronze				Graphite
Bells	80	20			
	Silicon bronze				
Blast furnace tuyeres	Copper				
and water jackets	Alloyed copper				
Carburetors	85	5	5	5	
	88	6.5	1.5	4	
	Yellow brass				
Cocks and faucets	85	5	5	5	
	81	3	7	9	
	76	2.5	6.5	15	
	Yellow brass				
Cutlery	Nickel silver				
	Aluminum bronze				
	Aluminum silicon bronze				
Dairy equipment	Nickel silver				
	Monel				
Dies, permanent mold	Copper				
	Alloyed copper				
	Aluminum bronze				
Dies, glass	Nickel aluminum bronze				
Dies, wire					
drawing	Rem.	9–11			1–4 Ni
	Aluminum bronze				
	Beryllium copper				
Electrical	Copper				
conductors	Chrome copper				
	Beryllium copper				
	93	1	1	5	
	85			15	
Fixtures,	85	5	5	5	
ornamental	81	3	7	9	
	Yellow brass				
	Nickel silver				
Flanges	85			15	
	85	5	5	5	
	88	6.5	1.5	4	

Table 10–2 Uses of copper alloys (*continued*)

Uses	Nominal compositions, %				
	Cu	Sn	Pb	Zn	Other
Fuel pumps	88	8		4	
	88	6.5	1.5	4	
	85	5	5	5	
Gears, general	85	5	5	5	
	88	6.5	1.5	4	
	Yellow brass				
	Manganese bronze				
	Aluminum bronze				
	Silicon bronze				
Gears,	89	11			0.01–0.5 P
heavy duty	87	11			2 Ni
	87	11	1		1 Ni
	88	10		2	
	Aluminum bronze				
	Manganese bronze				
Gear shifter forks	Aluminum bronze				
	Manganese bronze				
Impellers	85	5	5	5	
	88	6.5	1.5	4	
	88	8		4	
	78	7	15		
	Aluminum bronze				
	Manganese bronze				
	Silicon bronze				
	Silicon brass				
	Monel				
Injectors	88	8		4	
	88	6.5	1.5	4	
	85	5	5	5	
Jewelry	87	0–3		10–12	
	Yellow brass				
	Nickel silver				
Lamp fixtures	Yellow brass				
	Nickel silver				
Laundry machinery	Nickel silver				
	Monel				
	85	5	5	5	
Lock hardware	Yellow brass				
	85	5	5	5	
	Nickel silver				

Table 10–2 Uses of copper alloys (*continued*)

Uses	Nominal compositions, %				
	Cu	Sn	Pb	Zn	Other
Marine propellers	Manganese bronze				
	Manganese bronze				1–2.5 Ni
	Aluminum bronze				4–5 Ni
	12 % Manganese aluminum copper				2–3 Ni
	Monel				
Marine fittings	85	5	5	5	
and hardware	83	4	6	7	
	Silicon bronze				
	Manganese bronze				
	Monel				
	Aluminum bronze				
	Copper nickel				30 Ni
Memorial markers	88	1	3	8	
	88	6.5	1.5	4	
	Silicon bronze				
Meters,	85	5	5	5	
meter parts and	83	4	6	7	
gauge cases	Yellow brass				
Milking machine	Nickel silver				
parts	Monel				
Paper manufacturing	88	8		4	
machinery	88	6.5	1.5	4	
	85	5	5	5	
	80	10	10		
	78	7	15		
	Nickel silver				
	Aluminum bronze				
	Silicon bronze				
	Manganese bronze				
Pickling baskets	Aluminum bronze				
and tanks					
Piston rings, aircraft	82	18			
	80	16	4		
Piston rings, segmental	81	13	4.5		1.5 Ni
	85	5	5	5	
	88	8		4	
Packing rings	77	8	15		
for glands		To			
	65	5	30		

Table 10–2 Uses of copper alloys (*continued*)

Uses	Nominal compositions, %				
	Cu	Sn	Pb	Zn	Other
Pump bodies	85	5	5	5	
	88	6.5	1.5	4	
	88	8		4	
	80	10	10		
	78	7	15		
	Aluminum bronze				
	Silicon bronze				
	Nickel silver				
	Monel				
	70 Copper nickel				30 Ni
	90 Copper nickel				10 Ni
Railroad fittings	85	5	5	5	
	88	6.5	1.5	4	
	88	8		4	
	Nickel silver				
	Yellow brass				
Statuary and ornamental	Yellow brass				
	Nickel silver				
	81	2	2	15	
	85	5	5	5	
	89	1	3	7	
	88	6.5	1.5	4	
Superheated steam fittings	88	6.5	1.5	4	
	Monel				
Tools, non-sparking	Aluminum bronze				
	Beryllium copper				
Trolley shoes	Aluminum bronze				
	Silicon bronze				
Trolley wheels	88	8		4	
	93	5		2	
	Aluminum bronze				
	Silicon bronze				
Turbine nozzles	Monel				
	Nickel silver				
Valves, air compressor	88	6.5	1.5	4	
Valves, hydraulic	88	6.5	1.5	4	
	88	8		4	
	85	5	5	5	
	83	4	6	7	
	81	3	7	9	

Table 10–2 Uses of copper alloys (*continued*)

Uses	Cu	Sn	Pb	Zn	Other
	Nominal compositions, %				
Valves and fittings,	88	6.5	1.5	4	
steam	85	5	5	5	
	Aluminum bronze					
	Nickel silver					
	Monel					
Valve discs	88	8		4	
	88	6.5	1.5	4	
	85	5	5	5	
	Monel					
	Nickel silver					
Valve seats	52				(31 Ni, 12 Fe, 5 Cr)
	85	5	5	5	
	88	6.5	1.5	4	
	88	8		4	
	Aluminum bronze					
	Nickel silver					
	Monel					
Valve stems	Manganese bronze					
	Aluminum bronze					
	Silicon brass					
	Silicon-aluminum bronze					
	88	6.5	1.5	4	
	85	5	5	5	
	88	5		2	5 Ni
Wear plates	87	11			2 Ni
	Aluminum bronze					
Worm gears	87	11			2 Ni
and worm wheels	89	11			
	87	10	1	2	
	84	10	2.5		3.5 Ni
	Aluminum bronze					
	Manganese bronze					

Table 10–3 Uses of copper-base alloys (grouped according to specific alloys)

Alloys	Uses and applications	
High tin 10–20% Bronzes	Bearings: Bridge Electrical machinery Locomotive Machine tools Rolling mill Trunnion Turntable	Bells Gears Piston rings: Hydraulic sealing Pneumatic sealing Worm wheels
88–10–2 88–8–0–4	Bearings: Bridge Machine tool Trunnion Condensers Control manifolds Fuel pumps Gears Impellers Injecters Marine fittings	Paper machinery Pump bodies Railroad fittings Steam fittings Trolley wheels Valve: Bodies Discs Seats Stems
88–6.5–1.5–4	Bearings Carburetors Cocks and faucets Flanges Fuel pumps Impellers Injecters Paper machinery Pump bodies Railroad fittings Steam fittings	Statuary Valve: Discs Seats Steam Bodies Valves: Air compressor High pressure Hydraulic Steam
85–5–5–5	Architectural Bearings Carburetors Cocks and faucets Fixtures Flanges Fuel pumps Impellers Injecters Marine fittings Memorial markers Meters and parts	Paper machinery Pump bodies Statuary Valves: Hydraulic Pneumatic Stem Valve: Discs Seats Stems
81–3–7–9	Cocks and faucets Fixtures Flanges Marine fittings Ornamental	Statuary Valves: Hydraulic Pneumatic

Table 10–3 (*continued*)

Alloys	Uses and applications	
80–10–10 84–8–8 83–7–7–3	Bearings: Electrical machinery Machine tool Pump Railroad Rolling mill Self lubricating Trunnion	Impellers Paper machinery Pump bodies
77–8–15 to 65–5–30	Bearings: Diesel engine Gas engine Locomotive Pump Railroad Rolling mill Self lubricating	Gland seal rings and segments Impellers Paper machinery

Alloys	Uses and applications	
Yellow Brass	Architectural Band instruments Carburetors Cocks and faucets Die cast fittings Fixtures Lamp fixtures Locks	Medallions Meter parts Ornamental Paper machinery Railroad fittings Statuary Valves
Aluminum Bronze	Ball races Bearings: Machine tool Rolling mill Electrical cable clamps Gears and worm wheels Impellers Paper machinery Permanent molds Pickling baskets Structural components	Tools, non-sparking Trolley: Shoes Wheels Valve: Bodies Seats Stems Wear plates Worm wheels
Aluminum Silicon Bronze	Electrical cable clamps Gears Impellers Marine fittings Marine propellers	Pump housings Structural components Valve components Valve stems
Manganese Aluminum Copper	Marine components Marine propellers	

Table 10–3 (*continued*)

Alloys	Uses and applications	

Manganese Bronze	Architectural
	Ball races
	Bearings:
	  Rolling mill
	  Trunnion
	Gears
		Impellers
		Marine fittings
		Marine propellers
		Structural components
		Valve stems
		Worm wheels

Silicon Bronze	Bells
	Bridge bearings
	Impellers
	Marine fittings
	Memorial markers
	Paper machinery
		Pump bodies
		Trolley:
		  Shoes
		  Wheels
		Valves
		Valve stems

Silicon Brass	Bearings Electrical cable clamps Gears Impellers	Rocker arms Structural components Valve stems

Nickel Silvers	Architectural
	Bearings, turbine
	Dairy equipment
	Fixtures
	Locks
	Ornamental
		Paper machinery
		Railroad fittings
		Valve:
		  Bodies
		  Discs
		  Seats

Nickel Copper 70–30	Food machinery
	Impellers
	Laundry machinery
	Marine fittings
		Valve:
		  Bodies
		  Discs
		  Seats
		  Stems

Copper Nickel (70–30) (90–10)	Impellers
	Marine fittings
	Pump casings
		Valve:
		  Bodies
		  Discs
		  Seats
		  Stems

Copper Chromium 1% Heat-treated	Electrical conductivity:
	  Structural and switch components
	  Connectors	

Copper Beryllium (1–2%) Heat-treated	Electrical conductivity:
	  Structural and switch components
	  Contractors and connectors
	Dies—Wire drawing
	Non-sparking tools	

Table 10–4   Chemical composition limits[1] for aluminum ingot (aluminum alloy foundry ingot[2])

Alloy	Copper	Iron	Silicon	Manganese	Magnesium	Zinc	Nickel	Titanium	Tin	Chromium	Others Each	Others Total
12	6.0–8.0	1.2	1.0–4.0	0.50	0.07	1.1–2.5	0.30	0.20				0.50
13[3]	0.6	0.8	11.0–13.0	0.30	0.10	0.30	0.50		0.10			0.20
A13	0.10	0.6	11.5–12.5	0.05	0.03	0.05	0.05		0.05			0.10
43	0.10	0.6	4.5–6.0	0.10	0.05	0.10		0.20			0.05	0.15
108	3.5–4.5	0.8	2.5–3.5	0.30	0.03	0.20		0.20				0.30
A108	4.0–5.0	0.8	5.0–6.0	0.30	0.10	0.50		0.20				0.50
112	6.0–8.0	1.2	1.0	0.50	0.07	2.0	0.30	0.20				0.50
113	6.0–8.0	1.2	1.0–3.0	0.50	0.07	1.0–2.2	0.30	0.20				0.50
C113	6.0–8.0	1.2	3.0–4.0	0.50	0.07	2.5	0.50	0.20				0.50
122	9.2–10.8	1.2	1.0	0.50	0.15–0.35	0.50	0.30	0.20				0.30
A132	0.50–1.5	1.0	11.0–13.0	0.10	0.9–1.3	0.10	2.0–3.0	0.20			0.05	0.15
D132	2.0–4.0	1.0	8.5–10.5	0.50	0.50–1.5	0.50	0.50–1.5	0.20				0.50
138	9.5–10.5	1.2	3.5–4.5	0.50	0.15–0.35	0.50	0.50	0.20				0.50
142	3.5–4.5	0.6	0.6	0.30	1.3–1.8	0.10	1.7–2.3	0.20			0.05	0.15
A142	3.7–4.5	0.6	0.6	0.10	1.3–1.7	0.10	1.8–2.3	0.07–0.18		0.15–0.25	0.05	0.15
195	4.0–5.0	0.8	0.7–1.2	0.30	0.03	0.10		0.20			0.05	0.15
B195	4.0–5.0	0.8	2.0–3.0	0.30	0.03	0.30		0.20			0.05	0.15
212	7.0–8.5	1.2	1.0–1.5	0.30	0.05	0.20		0.20				0.30
214	0.10	0.30	0.30	0.30	3.7–4.5	0.10		0.20			0.05	0.15
A214	0.10	0.30	0.30	0.30	3.5–4.5	1.4–2.2		0.20			0.05	0.15
B214	0.10	0.30	1.4–2.2	0.30	3.6–4.5	0.10		0.20		0.20	0.05	0.15
C214	0.10	0.8	0.30	0.40–0.6	3.5–4.5	0.10					0.05	0.15
F214	0.10	0.30	0.30–0.7	0.30	3.5–4.5	0.10		0.20	0.10		0.05	0.15
L214	0.10	0.6–0.9	0.50–1.0	0.40–0.6	2.5–4.0	0.10	0.10		0.10		0.05	0.15
218	0.20	0.8	0.30	0.30	7.5–8.5	0.10	0.10					
218 Special	0.05	0.20	0.15	0.10–0.25	6.5–7.5			0.10			0.05	0.15
220	0.20	0.20	0.20	0.10	9.6–10.6	0.10		0.20			0.05	0.15

319(3)	3.0–4.5	1.0	5.5–7.0	0.8	0.50	1.0	0.50	0.20			0.50
A319	3.0–4.5	0.8	5.5–7.0	0.50	0.10	0.8	0.20	0.20			0.50
333	3.0–4.5	1.0	8.0–10.0	0.8	0.6	1.0	0.50	0.20			0.50
355	1.0–1.5	0.40(4)	4.5–5.5	0.30(4)	0.45–0.6	0.20		0.20		0.05	0.15
C355	1.0–1.5	0.15	4.5–5.5	0.10	0.45–0.60	0.10		0.20		0.03	0.10
356	0.20	0.40	6.5–7.5	0.10	0.25–0.40	0.10		0.20		0.05	0.15
A356	0.10	0.15	6.5–7.5	0.03	0.45–0.60	0.05		0.20		0.03	0.10
357	0.05	0.15	6.5–7.5	0.03	0.45–0.60	0.05		0.10–0.20		0.03	0.10
360(3)	0.6	0.8	9.0–10.0	0.30	0.40–0.6	0.35	0.50		0.10		0.20
A360	0.10	0.50	9.0–10.0	0.10	0.40–0.6	0.10				0.05	0.15
380(3)	3.0–4.0	1.0	7.5–9.5	0.50	0.10	0.9	0.50		0.30		0.50
A380	3.0–4.0	0.6	7.5–9.5	0.10	0.05	0.10	0.10			0.05	0.15
382	0.10	0.40	3.0–4.0	0.10	1.7–2.5			0.20	0.20	0.05	0.15
384	3.0–4.5	1.0	11.5–13.0	0.6	0.10	1.2	0.6		0.30		0.50
A612	0.35–0.65	0.4	0.15	0.05	0.6–0.8	6.0–7.0		0.20		0.05	0.15
C612	0.35–0.65	1.3	0.30	0.05	0.25–0.45	6.0–7.0		0.20		0.05	0.15
750	0.7–1.3	0.50	0.7	0.10			0.7–1.3	0.20	5.5–7.0		0.30

(1) Values are in percent maximum unless shown as a range—aluminum remainder.

(2) Reynolds chemical composition limits for aluminum alloy foundry ingot are designed to meet government and nongovernment specifications.

(3) The alloys A13, A319, C355, A356, A360, and A380 have the same compositions as alloys 13, 319, 355, 356, 360, and 380, respectively, but the impurities, notably iron and manganese, are controlled to closer limits.

(4) If the iron content exceeds 0.4%, the manganese content is approximately one-half the iron content.

Table 10–5 Typical mechanical properties of aluminum permanent-mold castings

Alloy	Tension Ultimate[2] strength, psi	Tension Yield strength (set 0.2%) psi	Tension Elongation (percent in 2 inches)	Compression Yield[3] strength (set 0.2%), psi	Hardness Brinell[2] 500-kg load, 10-mm ball	Shear Shearing strength, psi	Fatigue Endurance[1] limit, psi
43	23,000	9,000	10.0	9,000	45	16,000	
A108	28,000	16,000	2.0	17,000	70	22,000	
113	28,000	19,000	2.0	20,000	70	22,000	
C113	30,000	24,000	1.0	25,000	80	24,000	9,500
122–T52	35,000	31,000	1.0	31,000	100	25,000	
122–T551	37,000	35,000	(4)	40,000	115	30,000	8,500
122–T65	48,000	36,000	(4)	36,000	140	36,000	9,000
A132–T551	36,000	28,000	0.5	28,000	105	28,000	13,500
A132–T65	47,000	43,000	0.5	43,000	125	36,000	
138	32,000	24,000	1.5	32,000	100	22,000	
142–T571	40,000	34,000	1.0	34,000	105	26,000	10,500
142–T61	47,000	42,000	0.5	46,000	110	31,000	9,500
152–T524	29,000	16,000	1.0		95	22,000	
152–T74	35,000	26,000	0.5		100	29,000	
B195–T4[4]	37,000	19,000	9.0	20,000	75	30,000	9,500
B195–T6[5]	40,000	26,000	5.0	26,000	90	32,000	10,000
B195–T7	39,000	20,000	4.5	20,000	80	30,000	9,000
A214	27,000	16,000	7.0	17,000	60	22,000	
319	34,000	19,000	2.5	19,000	85	24,000	
319–T6	40,000	27,000	3.0		95		
333	34,000	19,000	2.0	19,000	90	27,000	14,500
333–T5	34,000	25,000	1.0	25,000	100	27,000	12,000
333–T6	42,000	30,000	1.5	30,000	105	33,000	15,000
333–T7	37,000	28,000	2.0	28,000	90	28,000	12,000
333–T533	32,000	25,000	1.0		100		
355–T51	30,000	24,000	2.0	24,000	75	24,000	

355–T6	43,000	27,000	4.0	27,000	90	34,000	10,000
355–T62	45,000	40,000	1.5	40,000	105	36,000	10,000
355–T7	40,000	30,000	2.0	30,000	85	30,000	10,000
355–T71	36,000	31,000	3.0	31,000	85	27,000	10,000
C355–T6	48,000	28,000	10.0		90		
356–T6	40,000	27,000	5.0	27,000	90	32,000	13,000
356–T7	33,000	24,000	5.0	24,000	70	25,000	11,000
A356–T6	41,000	28,000	12.0		80		
357–T6	49,000	36,000	8.0		85		
750–T533	20,000	8,500	10.0	8,500	45	13,000	9,000

(1) Endurance limits are based on 500,000,000 cycles of completely reversed stresses, using the rotating-beam type of machine and specimen.

(2) Tension and hardness values determined from standard half-inch-diameter tensile test specimens, individually cast in a permanent mold, or in green-sand molds and tested without machining off the surface.

(3) Results of tests on specimens having an 1/r ratio of 12.

(4) Less than 0.5%.

(5) When the casting stands at room temperature for several weeks, the tensile and yield strengths increase and elongation is reduced slightly. In the case of 195–T4, the properties approach those of the –T6 condition.

Table 10–6 Typical mechanical properties of aluminum die castings

	Tension			Shear	Fatigue
Alloy[3]	Ultimate[2] strength, psi	Yield strength (Set 0.2%), psi	Elongation, Percent in 2 inches	Shearing strength, psi	Endurance Limit[1], psi
13	39,000	21,000	2.0	25,000	19,000
A13	35,000	16,000	3.5		
43	30,000	16,000	9.0	19,000	17,000
85	40,000	24,000	5.0	26,000	22,000
L214	41,000		10.0		
218	45,000	27,000	8.0	27,000	23,000
360	44,000	27,000	3.0	28,000	19,000
A360	41,000	23,000	5.0	26,000	18,000
380	45,000	26,000	2.0	29,000	20,000
A380	46,000	25,000	3.0	29,000	19,000
384	46,000	27,000	1.0	29,000	21,000

(1) Endurance limits are based on 500,000,000 cycles of completely reversed stresses using the rotating-beam type of machine and specimen.
(2) Tensile properties are average values from ASTM standard round die-cast test specimen, $\frac{1}{4}$ inch in diameter, produced on a cold chamber (high-pressure) die-casting machine.
(3) The alloys whose numbers are prefixed by *A* differ from those without the prefix in that the impurities, notably iron, are controlled to lower limits.

Additions of copper which range from 2 to 5% improve the ductility of the aluminum alloy. Additions up to 12% increase its hardness and strength. However, the tendency toward hot-cracking gradually increases with the addition of copper up to 5%, although further additions decrease the alloy's susceptibility to hot-cracking.

Under equilibrium conditions, about 5.6% copper is soluble in aluminum at temperatures of solidification (1018°F or 547.89°C). At room temperature, the solubility is reduced to less than $\frac{1}{2}$% of copper. This decrease in solubility accounts for the susceptibility of aluminum-copper alloy to heat treatment. The mechanical properties of Al-Cu alloys are improved by heat-treatment of the solution and by age-hardening.

When silicon is added in amounts up to about 13%, there are no significant benefits from heat-treating. The properties of this Al-Si mixture are the results of alloying. Adding silicon increases strength at the expense of lessening ductility. Cooling quickly in a mold improves the properties of this alloy, and thus using metal molds enhances its properties. This alloy lends itself well to die-casting. Figure 10–9 shows the aluminum-copper equilibrium diagram.

The as-cast properties of Al-Si obtained through quick chilling or modification processes in the sand foundry are superior to those of most other aluminum alloys.

Table 10-7 Typical mechanical properties of aluminum sand castings

Alloy	Tension Ultimate[2] strength, psi	Tension Yield strength, (Set 0.2%) psi	Tension Elongation, percent in 2 inches	Compression Yield[3] strength (Set 0.2%) psi	Hardness Brinell[2] 500-kg load, 10-mm ball	Shear Shearing strength, psi	Fatigue Endurance[1] limit, psi
43	19,000	8,000	8.0	9,000	40	14,000	8,000
108	21,000	14,000	2.5	15,000	55	17,000	11,000
112	24,000	15,000	1.5	16,000	70	20,000	9,000
113	24,000	15,000	1.5	16,000	70	20,000	9,000
122–T2	27,000	20,000	1.0	20,000	80	21,000	9,500
122–T61	41,000	40,000	(4)	43,000	115	32,000	8,500
142–T21	27,000	18,000	1.0	18,000	70	21,000	6,500
142–T571	32,000	30,000	0.5	34,000	85	26,000	8,000
142–T77	30,000	23,000	2.0	24,000	75	24,000	10,500
195–T4	32,000	16,000	8.5	17,000	60	26,000	7,000
195–T6	36,000	24,000	5.0	25,000	75	30,000	7,500
195–T62	40,000	34,000	1.5	36,000	95	32,000	8,000
212	23,000	14,000	2.0	14,000	65	20,000	9,000
214	25,000	12,000	9.0	12,000	50	20,000	7,000
B214	20,000	13,000	2.0	14,000	50	17,000	
F214	21,000	12,000	3.0	13,000	50	17,000	
218 Special	36,000	18,000	9.0		65		
220–T4	46,000	25,000	14.0	26,000	75	33,000	10,000
319	27,000	18,000	2.0	19,000	70	22,000	8,000
319–T6	36,000	24,000	2.0	25,000	80	29,000	10,000
355–T51	28,000	23,000	1.5	24,000	65	22,000	10,000
355–T6	35,000	25,000	3.0	26,000	80	28,000	7,000
355–T61	39,000	35,000	1.0	37,000	90	31,000	9,000
355–T7	38,000	36,000	0.5	35,000	85	28,000	9,500
355–T71	35,000	29,000	1.5		75		10,000
C355–T6	39,000	29,000	5.0		80		10,000
356–T51	25,000	20,000	2.0	21,000	60	20,000	7,500
356–T6	33,000	24,000	3.5	25,000	70	26,000	8,500
356–T7	34,000	30,000	2.0	31,000	75	24,000	9,000
356–T71	28,000	21,000	3.5	22,000	60	20,000	
A356–T6	38,000	28,000	6.0		70		
357–T6	43,000	36,000	4.0		75		

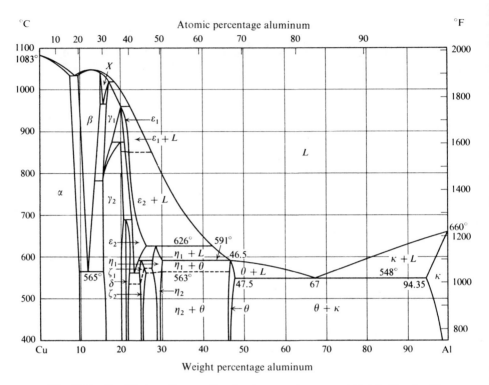

°C Atomic percentage aluminum °F
Weight percentage aluminum

Fig. 10–9. Equilibrium diagram for aluminum and copper. (*Metals Handbook*)

The solubility of silicon in aluminum at the solidification point (1070°F, 576.67°C) is 1.65%; this decreases to less than $\frac{1}{10}$ of 1% at room temperature. Alloys of aluminum and silicon are generally not heat-treated.

Aluminum alloyed with from 4 to 10% magnesium has properties similar to the properties it has when it is alloyed with copper. A ternary alloy of aluminum, magnesium, and silicon is amenable to solution and aging treatments, and thus this kind of an alloy can exhibit improved properties. In developing a quasi-binary alloy system, one must control the percentage of each element.

Aluminum-magnesium alloys are characterized by excellent mechanical properties, resistance to corrosion, and machinability. They have good resistance to impact, high ductility, and maintain their properties at elevated temperatures reasonably well. However, careful foundry handling is necessary because of their tendency to dross. These alloys have a narrow solidification range (see the equilibrium diagram in Fig. 10–10), and so feeding and chilling require special attention.

Magnesium up to about 15% is soluble in aluminum at the solidification temperature of the alloy (844°F or 441°C), although the solubility decreases to less than $2\frac{1}{2}$% at room temperature. Aluminum alloys with 6% or more of magnesium may be heat-treated.

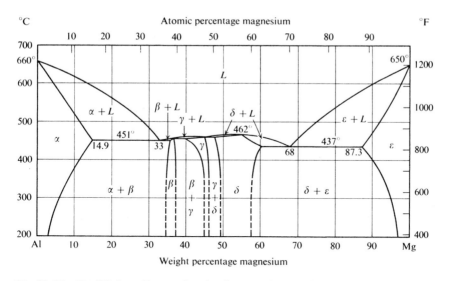

Fig. 10-10. Equilibrium diagram for aluminum and magnesium. (*Metals Handbook*)

When silicon is present, the compound magnesium silicide (Mg_2Si) is formed. The strength and hardness of this alloy are greater than the strength and hardness of Al-Mn, but there is a concomitant reduction of ductility. The presence of magnesium impairs the elongation and impact-resistance values in some alloys, and so the amount of magnesium must often be limited.

Some alloys, called *Tenzaloy* and *Ternaloy*, employ zinc as a principal alloying element. Tenzaloy has a zinc content of about 7.5% and Ternaloy about 3 to 4%. Zinc accounts for their excellent mechanical properties. These alloys reach optimum properties through self-aging at normal temperature in about two weeks. Tenzaloy has high tensile strength, yield strength, percent elongation, and good ductility properties, and obtains these without heat treatment. The normal composition of the alloy is 0·6% Cu, 7·5% Zn, 0·4% Mg, and the balance Al.

Due to casting difficulties and low strength, aluminum-manganese alloys are seldom used in the foundry. The resistance to corrosion and the ductility of these alloys, however, are excellent. Small amounts of manganese do not impair the mechanical properties of aluminum casting alloys.

Iron is a natural element to be alloyed with aluminum, since it exists with aluminum in bauxite ore. Although iron is considered an impurity, it performs a useful function. Small percentages of iron increase the strength and hardness of certain alloys and reduce the tendency to hot-cracking.

However, mechanical properties of an alloy are progressively impaired as the iron constituent increases in percentage. Rapid chilling keeps the size of the iron constituent to a minimum. Alloys cast in dies or permanent molds are usually permitted a higher iron content than alloys cast in sand molds.

Heat treatment of an aluminum casting is a process in which the casting is heated and then cooled under precisely regulated conditions for the following purposes.

1.   Developing a uniform structure throughout the casting.

2.   Removing internal stresses caused by thermal conditions or contraction during solidification.

3.   Improving mechanical properties.

4.   Improving dimensional stability.

The heat treatment of aluminum castings usually involves two distinct operations: *solution treatment* and *artificial aging*. They are performed independently of each other. Either one or both of the operations can be used, depending on the results desired.

To obtain a casting with a homogeneous structure, you must use solution treating or redissolving. This involves heating the casting to a temperature in the range of 820°F (437.8°C) to 1000°F (537.8°C) at which the precipitated elements will, for the most part, redissolve. The temperature varies with the chemical compositions of the alloy. It must be somewhat below the solidus temperature of the alloy, since at or beyond this point the casting might even melt. Since atomic mobility increases with temperature, the precipitated elements are redissolved at the highest temperature that will not harm the casting.

A coarser structure requires a longer solution-treating time than a fine one. It is important to allow enough time so that the casting is at a completely uniform temperature. A period of 12 to 15 hours is considered sufficient to produce the necessary changes.

Once the structure has changed, you must retain the structure the casting has at the solution-treating temperature. This is done by cooling it quickly by quenching in near-boiling water. If it is cooled by quenching in cold water, distortion or cracking may result. It is most important to avoid delay in quenching. Since the transfer of heat from the casting occurs only at its surface, the casting must be completely submerged in the water.

The alloy is in an unstable condition at this period. Any straightening or correction for distortion is best carried out within 24 hours, while the metal is in a comparatively ductile condition.

The basis of the aging treatment is the unstable solid-solution structure developed by the solution treatment and subsequent quenching. Some elements begin to precipitate due to the unstable condition of the solution-treated alloy. The change may take from several weeks to several months to reach completion. Artificial aging is used to hasten this process (since re-precipitation occurs more rapidly at higher temperatures) so that the casting can be placed in service at an earlier period.

Artificial aging is conducted at a lower temperature than solution treatment, usually about 300°F (148.9°C). Artificial aging is a function of time and temperature, so that results may be controlled by varying either of these two factors. Too high a temperature causes overaging and worsens mechanical properties.

Artificial aging usually increases the strength and hardness of an alloy, but lessens its ductility.

You can relieve stress within the casting by heating it in the range of 450°F (232.2°C) to 550°F (287.8°C) and then air-cooling it. This treatment stabilizes the alloy so that further dimensional changes will not occur at lower temperatures.

When differing metals are electrically connected, as when both are immersed in brine, normal rates of corrosion are altered. The corrosion of one metal is accelerated, while that of the other is retarded. This is described as *galvanic corrosion*, and comes about when two dissimilar metals are in contact in an electrolyte, which causes them to generate a galvanic or direct current.

The metal with the greatest corrosion is said to be the *anode* or "less-noble" metal; the metal with the lesser corrosion is the *cathode* or "more-noble" metal. The common dry-cell battery utilizes galvanic reactions. The zinc anode undergoes corrosion, while the copper or carbon cathode is quite unaffected. This same principle is utilized in the protection of ship hulls and engines. Zinc or magnesium anodes, which are sacrificed through electrolysis action, are strategically placed on the hull and on the engine.

Table 10–8 lists the activity in galvanic corrosion of various metals and alloys.

Table 10–8 Relative activity of various metals in galvanic corrosion

Corroded end (anodic or less-noble)
    Magnesium
    Aluminum
    Zinc
    Cadmium
        Steel, cast iron
        Stainless steels (active)
            Soft solders
            Tin
            Lead
                Nickel
                Brasses
                Bronzes
                Monel
                Copper
                    Stainless steels (passive)
                    Silver solder
                      Silver
                      Gold
                      Platinum
                    Protected end (cathodic or most-noble)

Utilizing conventional foundry processes in a selective manner so as to produce a cast product with the most desirable characteristics of each process produces a premium-quality casting, since a combination of features not found in any one method is used to produce these castings. The technique has developed to meet the exacting demands of the electronics and the aerospace industries for quality and reliability.

The precision or accuracy of premium-quality castings depends largely on the care with which models and patterns are made. Mold materials were developed which would ensure close dimensional accuracy and smooth surface finish, and would make possible the maximum amount of precision. Plaster or metal molds are often used to produce close dimensional tolerances and smooth surface finish. Expendable wax or plastic patterns are used to produce intricate parts.

The most important alloys used to make castings are those of the aluminum-silicon-magnesium and aluminum-copper-magnesium systems.

Mechanical properties in premium-quality castings are related directly to the solidification pattern achieved by the foundry. Thus the principles of heat-treating are important in attaining optimum mechanical properties.

Structural forms for use in aerospace vehicles are produced as premium-quality castings. These castings have tensile strength as high as 57,000 psi, with yield strength of 46,000 psi and elongation capability of 3 %.

BIBLIOGRAPHY

1. American Foundrymen's Society, *Copper-Base Alloys: Foundry Practice*, third edition, Des Plaines, Ill., 1965
2. *Copper and Copper Alloys*, Copper Development Association, London, 1962
3. R. W. Heine, C. R. Loper, Jr., and P. C. Rosenthal, *Principles of Metal Casting*, McGraw-Hill, second edition, New York, 1967
4. *Brass and Bronze Casting Alloys*, American Smelting and Refining Company, New York, 1967
5. American Society for Metals, *Metals Handbook*, eighth edition, Vol. 1, 1961
6. American Society for Metals, *Aluminum*, Vol. 3, Metals Park, Ohio, 1967
7. *Federated Aluminum Casting Alloys Handbook*, American Smelting and Refining Company, New York, 1966

CHAPTER 11

FERROUS METALS:
THEIR PROPERTIES AND USES

Cast metals are generally classified in two groups: *ferrous* and *nonferrous*. Ferrous metals, which may be subdivided according to carbon content and classed as steel or cast iron, are discussed in this chapter. Nonferrous metals were discussed in Chapter 10.

GRAY CAST IRON

Gray iron—commonly known as *cast iron*—is more widely used than any other casting metal or alloy. It is defined as an alloy of iron, carbon, and silicon, in

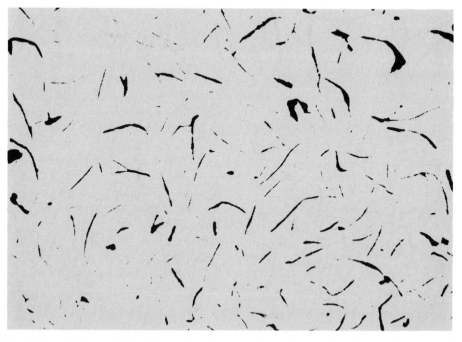

Fig. 11–1. Flake graphite in thin section of gray cast iron of approximately eutectic composition (unetched); 100 × approximately. (Courtesy A. P. Gagnebin – International Nickel.)

(a) Longest flakes 4 in. or more in length

(b) Longest flakes 2 to 4 in. in length

(c) Longest flakes 1 to 2 in. in length

(d) Longest flakes $\frac{1}{2}$ to 1 in. in length

(e) Longest flakes $\frac{1}{4}$ to $\frac{1}{2}$ in. in length

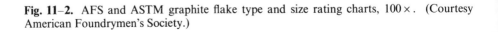

(f) Longest flakes $\frac{1}{8}$ to $\frac{1}{4}$ in. in length

Fig. 11–2. AFS and ASTM graphite flake type and size rating charts, 100×. (Courtesy American Foundrymen's Society.)

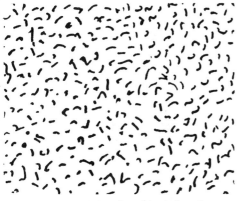

(g) Longest flakes $\frac{1}{16}$ to $\frac{1}{8}$ in. in length

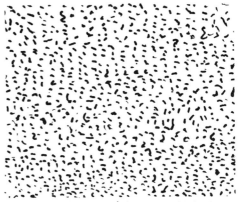

(h) Longest flakes $\frac{1}{16}$ in. or less in length

(i) Uniform distribution,
random orientation

(j) Rosette groupings

(k) Superimposed flake sizes,
random orientation

(l) Interdendritic segregation,
random orientation

(m) Interdendritic segregation,
preferred orientation

which the carbon content is greater than 2%. When the carbon content is less than this amount, the alloy is classified as steel. Cast iron contains free graphite or carbon, whereas steel does not. In general, the carbon content of cast iron is about 3.6%, although this figure may vary from between 2.5 to 4.0% by weight. Lesser amounts of carbon, down to 2.5% produce iron with greater strength.

In addition to iron, carbon, and silicon, gray iron may also contain small percentages of manganese, sulfur, and phosphorus, which may be present as a result of the blast-furnace process. When alloy cast irons are desired, nickel, chromium, and molybdenum may be added.

A characteristic of gray cast iron is that a large portion of its carbon exists in the form of graphite flakes (free carbon), as in Fig. 11–1. Its low specific gravity accounts for the fact that gray cast iron occupies more than three times as much space as an equal weight of metal.

The properties and uses of cast irons are largely related to the two alloying elements, carbon and silicon, and their effects on the process of graphitization. As the percentage of these elements in the iron increases, the formation of graphite is increased, and the graphite forms into various shapes (Fig. 11–2). Carbon may occur in cast irons as a chemical compound (Fe_3C); or it may be in the form of cementite. In these cases it is referred to as *combined carbon*. *Graphitization* is the process by which free carbon (graphite) is precipitated in the iron. Iron and carbon combine chemically to form Fe_3C, and it is this carbon that changes to free carbon. Graphitization is enhanced when the carbon content is greater than 2.0%. Silicon, by causing the iron carbide to become less stable, promotes the formation of graphite.

WHITE CAST IRON

Disregarding cooling rates and any other variables, silicon in amounts of less than 1.5% contributes to the formation of white iron. When the percentage of silicon is increased, the iron is changed from white to mottled or gray. The transitional area between the gray and the white iron is called *mottled* iron. Thus carbon and silicon may be varied to produce a white or gray iron, as desired. White iron may be produced in one of two ways: (1) By cooling the casting rapidly enough to prevent the removal of carbon from chemical combination. (2) By proper adjustment of the composition.

Because of rapid loss of heat, those castings which contain thin sections have a greater tendency to becoming white iron than those containing thick sections. Another influential factor is the heat conductivity of the mold material. A metal mold is more likely to produce a casting of white structure than a green-sand mold. If one inserts a piece of metal in a given area of a mold, this acts as a chill and produces a finer-grained structure in this area of the casting.

Under certain conditions, portions of a casting may be completely white and other portions completely gray. Castings of this type are called *chilled castings;* they are usually hard, and have wear-resistant surfaces. The depth of chill in such an iron varies with the percentages of carbon, silicon, manganese, phosphorus, and sulfur present in the metal. The rate of cooling of a casting affects the degree to which white iron is formed (Fig. 11–3). A casting that is cooled quickly produces the depth of chill necessary to create white iron castings.

MALLEABLE IRON

White irons, which are used for making malleable iron castings, contain less carbon and silicon than gray irons, and the carbon is in a combined form, as-cast. The carbides in white cast iron are unstable. When they are in a solid state they can be slowly graphitized by means of heat treatment. This produces malleable iron (Fig. 11–4).

When the carbon content of a metal is increased to the point where some of the carbon is in the form of free carbon, the strength and hardness of the metal are

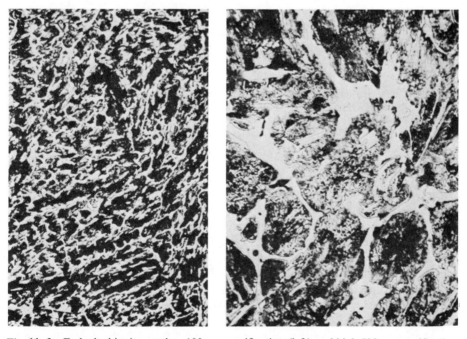

Fig. 11–3. Etched white iron at low 100 × magnification (left) and high 500 × magnification (right). The white areas are spines of carbon in the combined form, called cementite or Fe_3C, a very hard and brittle constituent. The matrix is substantially fine pearlite. (Courtesy Malleable Iron Foundry Society.)

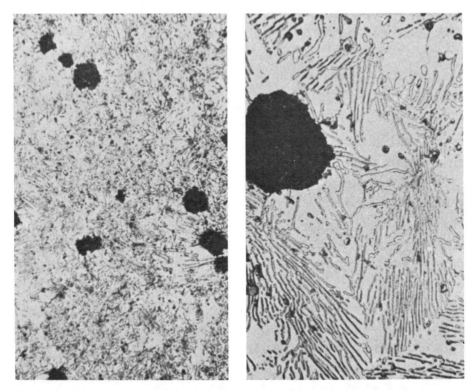

Fig. 11–4. Microstructure of malleable iron with insufficient manganese to balance the sulfur content. Ratio of manganese to sulfur was approximately 1:1. This sample was given a standard anneal. Note the spherulitic temper carbon and the retained coarse pearlite. Nital etch; 100× (left) and 500× (right). (Courtesy Malleable Iron Foundry Society.)

reduced. The free carbon or graphite breaks up the continuity of the metallic structure. At the same time it acts as a lubricant, thus increasing the machinability of the metal, as well as its capacity to damp vibrations.

The presence of silicon in cast iron is the most important single composition factor promoting graphitization in gray cast iron. Because the presence of silicon lessens the ability of iron to dissolve carbon, an increase in silicon reduces the depth of the chill. The addition of a sufficient quantity of silicon completely eliminates white chill. Table 11–1 lists typical chemical compositions of a few commercial cast irons, and Table 11–2 gives a simplified comparison of the properties of cast ferrous alloys.

The amount of silicon in iron may vary from 1.0 to 3.75%, depending on the type of metal desired.

Sulfur acts as a carbide stabilizer, but is considered harmful in amounts greater than 0.15%. Excess sulfur reduces fluidity, increases shrinkage, produces

Table 11-1 Typical chemical composition of some commercial cast irons*

Iron	C, %	Combined carbon, %	Si, %	Mn, %	S, %	P, %	Other, %	Bhn	Min. tensile strength, psi	Use
Gray	3.30–3.60		2.30–2.60	0.50–0.80	0.20 max	0.30 max		192 max	30,000	General-purpose use Motor blocks
	3.10–3.50	0.40–0.70	1.90–2.30	0.60–0.90	0.125 max	0.12–0.18		163–228		Piston rings
	3.50–3.90		2.20–3.10	0.40–0.80	0.10	0.30–0.80		222–267		
	2.90–3.20	0.65–0.90	0.90–1.10	0.65–0.90	0.05–0.12	0.20 max	1.00–1.50 Ni 0.50 Cr	200–240	40,000	Heavy machine-tool bases, 2000–10,000 lb
	2.60–2.80	0.60–0.75	2.20–2.50	0.90–1.00	0.08 max	0.08 max	0.75–1.00 Ni 0.75–1.25 Mo 0.10–0.20 Cr	220–240	60,000	High-strength iron, large diesel-engine crankshafts
Chilled	3.25–3.60		0.50–0.65	0.40–0.60	0.15 max	0.30–0.45		45 RC as-cast		Chilled-iron freight-car-wheels, rolls
White, malleable	2.20–2.40	†	0.90–1.10	0.35–0.50	0.12 max	0.14 max	0.03 Cr max	Over 320,‡ 135 max§	50,000	General-purpose malleable iron
Cupola, malleable	2.70–3.20	†	0.60–0.80	0.45–0.60	0.15 max	0.15 max				Malleable-iron pipe fittings
Nodular	3.60–4.20	0–0.20	1.25–2.00	0.35		0.08	0–1.0 Ni 0.05–0.08 Mg	140–200	60,000	Pressure castings, valve and pump bodies, shock-resisting parts
	3.20–3.80	0.70	2.25–2.75	0.60–0.80		0.10	1.5–3.5 Ni 0.05–0.08 Mg	200–270	80,000	Heavy-duty machinery, gears, dies, rolls, for wear resistance and strength

* Adapted from *Cast Metals Handbook*.
† Virtually all the carbon is present as carbide in the white iron, but is graphitized by heat treatment.
‡ Before malleableizing heat-treatment.
§ After malleableizing heat-treatment.

Table 11-2 Comparison of cast ferrous alloys

Alloy class	Approximate analysis, %	Melting range, °F	Compression, psi	Tensile strength, psi	Elonga-tion, %	Yield, psi	Modulus of elasticity, psi	Remarks
Gray iron	Total C, 2.5–3.5 Si, 1–3.25 Mn, 0.5–0.75 P, 0.15–1 S, max 0.10	2100–2300	105,000–175,000	25,000–60,000	<1	Does not exhibit definite values	14,000,000–20,000,000	Total C as flake graphite Good machinability Good damping High wear resistance Not very tough Low tensile strength; not malleable
White iron	Total C, 1.7–2.5 Si, 0.85–1.2 Mn, 0.25–0.80 P, 0.06–0.20 S, 0.06–0.18	2300–2500		50,000–60,000	<1	Same as for gray iron	Same as for gray iron	Total C as cementite Hard to machine Abrasion-resistant
Malleable iron	Total C, 1.5–2.3 Si, 0.85–1.2 Mn, max 0.40 P, max 0.20	2600–2800	90,000+	50,000–60,000	10–25	35,000–40,000	25,000,000–30,000,000	Cast as white iron, annealed by heating to 1700–1800°F Very good machinability
Nodular iron	Total C, 3.2–3.8 Si, 1–3.25 Mn, 0.35–0.80 P, 0.08–0.10 S, 0.008–0.01	2100–2300	120,000–200,000	55,000–120,000	≦18	40,000–90,000		Total C as spheroids Properties as cast but may be annealed Maximum ductility and toughness

	Carbon steel	Alloy steel
Composition	Total C, 0.1–1.4 Si, 0.2–0.7 Mn, 0.5–1 P, max 0.05	Total C, 0.2–0.5 One or more of following (minimum %): Mn, 1; Si, 0.7; Ni, Cu, 0.5; Cr, 0.25; Mo, 0.1; others, 0.05
	2700–3000	2700–3000
	65,000–130,000	70,000–200,000
	64,000–130,000	5–35
	20–35	45,000–170,000
	35,000–75,000	Same as for carbon steel
	~30,000,000	
Remarks	Heat-treated after casting to provide desired properties High strength High toughness Machinability good to poor	A wide variety of analyses makes properties hard to generalize

chill, and makes iron hard and brittle. To counteract the ill effects of a given amount of sulfur requires from 10 to 15 times the amount of silicon.

Manganese, when added in quantities in excess of that required to neutralize sulfur, is a carbide stabilizer. Since sulfur has a greater affinity for manganese than it does for iron, it forms manganese sulfide, which is insoluble in iron and which exists as a relatively harmless metalloid. Some of this manganese sulfide can be skimmed off with the slag, as it floats to the surface. If 0.30% more manganese is present than is needed for chemical combination with sulfur, many of the bad effects of sulfur may be eliminated. In commercial cast iron, between 0.50 and 0.80% manganese is the normal allowable amount.

Phosphorus unites with iron to form iron phosphide, and has little effect on the graphitization of carbon. The eutectic—consisting of iron, iron phosphide, and cementite—is called *steadite*. The addition of phosphorus increases the hardness of iron, as well as improving fluidity by reducing the melting temperature. Commercial iron contains from 0.15 to 0.90% phosphorus.

Malleable iron is a desirable engineering material because of its ease of machinability, its toughness, ductility, and wide range of strengths. Some of the principal industries using castings made of malleable iron are: the automotive, railroad, agricultural machinery, marine and mining equipment, and pipe fittings industries.

Malleable iron is a ferrous alloy composed of ferrite, which contains dissolved silicon and temper carbon (see Glossary of Terms). The carbon–silicon ratio must be such that it produces a matrix of ferrite and cementite. The final structure is the result of the heat treatment of white iron. Table 11–3 gives the chemical composition of common grades of white iron that are heat-treatable to produce malleable iron.

Table 11–3 Chemical composition of common grades of iron

Element	Gray iron, %	White iron* (malleable iron), %	High-strength gray iron, %	Nodular iron,† %
Carbon	2.5–4.0	1.8–3.6	2.8–3.3	3.0–4.0
Silicon	1.0–3.0	0.5–1.9	1.4–2.0	1.8–2.8
Manganese	0.40–1.0	0.25–0.80	0.5–0.8	0.15–0.90
Sulfur	0.05–0.25	0.06–0.20	0.12 max	0.03 max
Phosphorus	0.05–1.0	0.06–0.18	0.15 max	0.10 max

* Such compositions may be converted from white to malleable iron by heat treatment.
† Necessary chemistry also includes 0.01 to 0.10% Mg.

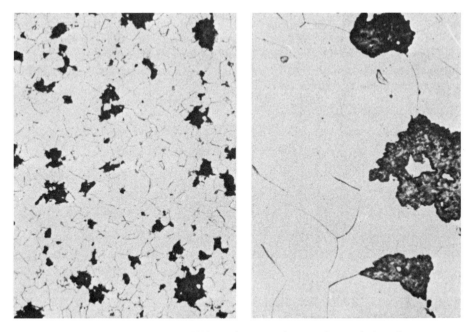

Fig. 11–5. Ferritic malleable iron exhibits only two major constituents in its microstructure: ferrite and temper carbon. The dark particles are temper carbon, while the white background material represents the ferritic matrix. Grain boundaries are clearly visible. The small gray particles on the right are manganese sulfide. The etchant was nital. Magnifications, 100 × (left) and 500 × (right). (Courtesy Malleable Iron Foundry Society.)

Figures 11–3 and 11–5 show structures of white iron and malleable iron. The white-iron castings are packed in annealing boxes containing slag, sand, or gravel; they are sealed, placed in a furnace, and heated for about 20 hours until the castings acquire a temperature of 1600°F (871°C). They are held at this temperature for a period of 40 hours. The temperature is then lowered at a rate of 10 degrees per hour until it reaches 1275°F (690°C), at which time the furnace is opened and the castings are allowed to cool to room temperature. In this heating process, the white iron, consisting of free cementite in a matrix of pearlite, is transformed to temper carbon in the matrix of ferrite. The temper carbon appears as free carbon in the form of nodules. In this form, temper carbon considerably increases the strength and ductility of the iron. The slow heating and cooling process provides a casting relatively free of internal stress. The absence of cementite, together with the lubricating action of carbon, enhances the machinability of malleable iron. Tensile strengths in the area of 50,000 psi and yield strengths at 65 % of the ultimate strength are common properties of malleable iron. By varying the heat-treating process, one may make malleable iron harder, stronger, and more resistant to wear, although one may lessen the iron's ductility and resistance to shock. When the transformation from combined carbon to nodular carbon is incomplete, the resultant type of iron becomes known as pearlitic malleable iron.

Figure 11–6 illustrates this overall composition range with respect to carbon and silicon in cast irons. Gray irons and white irons have a number of alloys falling within broad limits of composition.

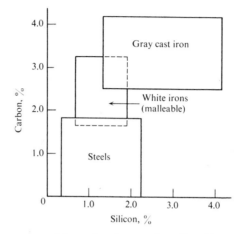

Fig. 11–6. Overlapping compositions of various grades of iron illustrated graphically. (From Heine, Rosenthal, and Loper.)

DUCTILE IRON

A relatively new kind of iron which is growing in importance is ductile iron. Discovered in 1947, it is an iron that in its as-cast original contains graphite which grows as spherulites rather than as flakes (Fig. 11–7). To produce this, one treats the molten iron with a small percentage of magnesium, cerium, or other agent that will cause a large proportion of its carbon to occur as spheroids of graphite. A base iron with a sulfur content consistently less than 0.025 % is desired in order to obtain an effective residual amount of magnesium from batch to batch.

To obtain a dependable carbide-stabilizing effect from the magnesium, one needs a balance between sulfur and magnesium. Ferro-silicon used as a post inoculant greatly increases the nodule count in castings. For maximum effectiveness, some ferro-silicon should be added at every transfer of metal, that is, at the bull ladle, pouring ladle, and the sprue.

To maintain high nodule counts and low amounts of carbide, it is necessary to have a high pouring temperature, high carbon equivalent, and post inoculant. This produces ductile iron of a high quality. In general, high-quality ductile iron requires the initial development of a high nodule count and a processing cycle which maintains that count until the casting is solidified.

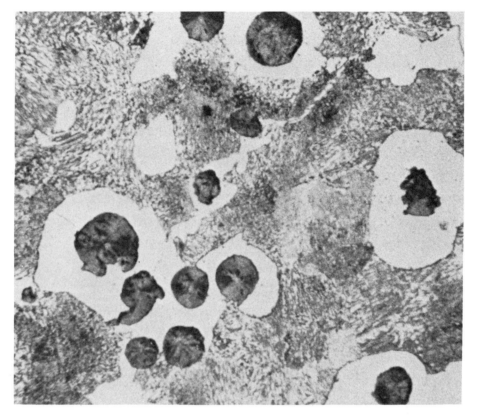

Fig. 11–7. Ductile cast iron. Note the graphite in nodular or spherulitic form, surrounded by "bull's eyes" of ferrite. The matrix is pearlitic. Nital etch; magnification 500 ×. (Courtesy Malleable Iron Foundry Society.)

The structure of the matrix has the greatest effect on the properties of the iron. The presence of graphite spheriods has only a minor influence on the mechanical properties. The matrix of ductile iron can be controlled by altering the base composition or by varying foundry practices. Ductile iron can be heat-treated to produce a tensile strength of 60,000 psi and an elongation of 25 %. Or the matrix can be controlled so that the ductile iron has a tensile strength of up to 150,000 psi with up to 4 % elongation. Table 11–4 illustrates some of the types of ductile iron.

It is during the solidification process that the differences between gray, mottled, and chilled iron are established. Figure 11–8 is a simplified schematic diagram illustrating their relationship.

Table 11-4 Principal types of ductile iron*

Type no.†	Brinell hardness no.	Characteristics	Applications
80–60–03	200–270	Essentially pearlitic matrix, high strength as-cast; responds readily to flame or induction hardening	Heavy-duty machinery, gears, dies, rolls for which wear resistance and strength are needed
60–45–10	140–200	Essentially ferritic matrix, excellent machinability and good ductility	Pressure castings, valve and pump bodies, shock-resisting parts
60–40–15	140–190	Fully ferritic matrix, maximum ductility and low transition temperature (has analysis limitations)	Navy ships and other environments requiring shock resistance
100–70–03	240–300	Uniformly fine pearlitic matrix, normalized and tempered or alloyed; excellent combination of strength, wear resistance, and ductility	Pinions, gears, crank-shafts, cams, guides, track rollers
120–90–02	270–350	Matrix of tempered martensite; may be alloyed to provide hardenability; maximum strength and wear resistance	

* Courtesy Gray and Ductile Iron Founders' Society.
† The type numbers indicate the minimum tensile strength, yield strength, and percent of elongation. The 80–60–03 type has a minimum of 80,000 psi tensile strength, 60,000 psi yield, and 3 per cent elongation in 2 in.

a) Liquid metal cools until freezing or solidifying begins at 1 (Fig. 11–8). At this point solid austenite dendrites begin to form and grow, and continue until the upper temperature in region 2 is reached.

b) Eutectic (a *eutectic* is a liquid saturated with respect to two solids) freezing begins as the temperature falls and area 2 is entered. The eutectic solids which form may be a mixture of austenite and carbide (which means that white iron is being formed) or of austenite and graphite (which means that gray iron is being formed). Graphite will predominate if graphitizing factors, such as high silicon content and

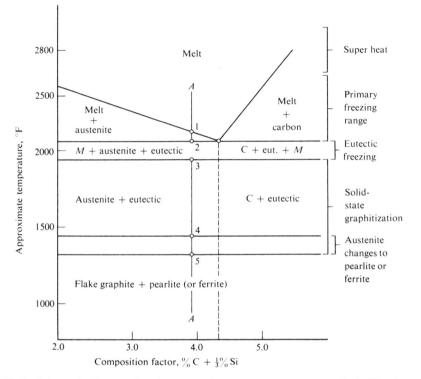

Fig. 11–8. Schematic diagram showing approximate temperature range of solidification and graphitization in cast irons.

slow cooling rate, are operative. Low silicon content and rapid cooling cause the eutectic to freeze as a mixture of carbide and austenite (white). When the temperature drops to the lower part of area 2, freezing is completed. Thus an iron freezes as white or gray iron. If it freezes gray, the nature of the graphite is established during freezing. Mottled irons are borderline cases in which both freezing mechanisms have occurred.

c) At the end of the freezing process, the structure consists of solids developed during steps (a) and (b). In gray irons these are austenite and graphite, and in white irons, austenite and carbide.

d) Further cooling in area 3 results in the precipitation of carbon from the austenite present, since the austenite may contain as much as 2.0 % C at the end of freezing, but only about 0.60 to 0.80 % as the temperature decreases to the temperature of area 4. The excess of carbon in the austenite is precipitated as carbide in white irons and as graphite in gray irons.

e) In area 4, the final change occurs in the solid state during cooling. Austenite transforms over the temperature range of areas 4 to 5. Because this change is quite complex, one can make only a few generalizations. With the most favorable of graphitizing conditions, only ferrite is formed in gray irons. With less-severe graphitizing conditions, ferrite and pearlite, or pearlite only, are formed. In white irons, only pearlite is formed. The less-severe conditions lead to a final microstructure of white iron such as is used for malleable castings.

f) Cooling below the temperatures in area 5 to room temperature produces little change in the iron.

Obviously the type of iron—white, mottled, chilled, or gray—is largely established during the freezing process. An iron's microstructure at room temperature reflects the entire freezing and cooling process of the iron. Thus the properties of cast irons are greatly influenced by the thermal and chemical changes occurring during its entire history, from liquid metal to cooled casting.

EFFECTS OF ALLOYING ELEMENTS

Small quantities of nickel added to iron produce fine-grained crystal structures and prevent segregation (see Glossary of Terms). Nickel is soluble in iron in both the liquid and solid state, and has a distinct advantage in heat-treating, as it lowers the transformation temperature and retards grain growth at elevated temperatures. With the lower temperature, cracking due to metal expansion is reduced, and a wider temperature range is allowable for heat-treating.

Nickel alloyed to cast iron improves the iron's machinability, strength, and resistance to corrosion. Since nickel is a graphite-forming element, it prevents low-carbon cast iron from forming white iron. Thus it also retards chills in thin sections. Additions of nickel to cast iron vary from 0.25 to 5.0% depending on requirements. More than 5% makes the iron extremely hard.

When nickel is added to steel, there is a decided improvement in the steel's tensile, impact, and fatigue strengths. Additions of nickel also improve steel's resistance to corrosion in fresh water, salt water, many acids, salts, and alkalies. The fact that nickel, when added to steel, lends improved strength and greater heat resistance at elevated temperatures makes nickel a reliable addition to components used in space exploration. Excavating and mining machinery as well as rolling mills and marine equipment also make use of nickel-alloyed iron and steel because of the improved properties resulting from the alloying with nickel.

When iron and steel are alloyed with chromium, molybdenum, silicon, and other elements, many desirable properties are obtained.

Chromium is one of the most valuable elements to alloy with iron and steel when one wishes to obtain resistance to corrosion and heat. Nickel and chromium are employed as alloys in steel used to make components of nuclear energy installations. These alloys are used in proportions ranging from a very small percentage up to 35%, depending on the properties desired. Chromium, when added to iron or steel, prevents segregation, thus producing a more uniform structure. The magnetic properties of permanent magnets are improved when 5% chromium is added. Steels containing 12% or more of chromium are in the stainless steel class. The strength of cast iron is also improved by additions of chromium which reduce the flake size of graphite, and also retard the development of a pearlitic structure of the metal. Cast iron with a high chromium content has unusual resistance to abrasion.

Molybdenum added to steel helps to reduce grain growth and leads to a fine-grained pearlitic structure, since molybdenum inhibits the decomposition of austenite.

Manganese added to steel is a powerful deoxidizer; it also renders sulfur inactive. It produces steel which has a fine grain structure, high strength and improved ductility. Steels with a medium amount of manganese contain 1 to 3% manganese, and have a pearlitic structure of fairly high strength and medium hardening properties. Some steels range up to 10 to 14% manganese; these are used in special applications. Up to 1 per cent manganese in iron does not affect other properties; amounts in excess of 1.25% produce marked increases in strength, hardness, and depth of chill.

Alloys containing less than 4.3% carbon (but more than 1.7%) are known as *hypoeutectic alloys*. Those containing more than 4.3% carbon are called *hypereutectic*. If additional carbon is added to iron in the iron carbide system, the proportion of iron carbide gradually increases until a concentration of 6.7% carbon is reached, at which time the entire structure is iron carbide, the compound Fe_3C.

In a hypoeutectic iron with 3% carbon, solidification commences at about 2340°F (1282°C). Austenite separates from the melt, and, being lower in carbon, it has the effect of increasing the carbon content in the remaining liquid. This continues as the temperature falls, until it reaches about 2065°F (1240°C). The material then consists of solid austenite containing about 1.7% carbon and liquid iron containing about 4.3% carbon. The remaining liquid freezes in the eutectic form, the mixture of iron carbide and austenite known as *ledeburite*.

Figure 11–9 shows an example of hypoeutectic iron. The islands were formerly the austenite crystals which separated from the melt. The speckled structure is the ledeburite or eutectic that froze at 2065°F (1240°C). With further cooling, the austenite steadily rejects iron carbide until it reaches a concentration of 0.83% carbon at a temperature of 1333°F (722°C), where it transforms to pearlite. The large and small islands that separate out when the eutectic freezes are transformed to pearlite. Figure 11–10 shows the graphite structure in hypereutectic iron.

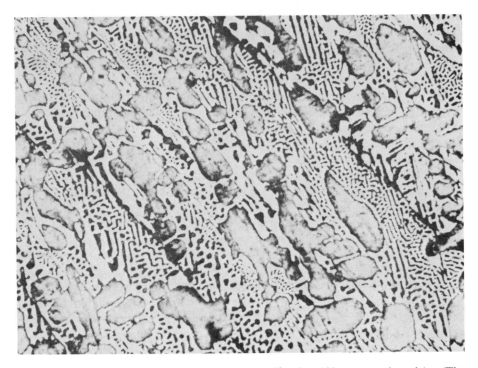

Fig. 11–9. Hypoeutectic white iron (etched; magnification 100 × approximately). (The British Cast Iron Research Association)

The properties of steel castings are uniform regardless of the direction in which the castings are tested. However, structural shapes produced by working down ingots or billets do develop directional properties during the working. Thus cast steel, which does not have this directionality, is better suited in many applications. Steel castings may be easily welded with no serious changes in their properties. Because of this, there has evolved the *cast-weld process,* in which large or irregular shapes are cast individually and then welded together to make a complex steel shape.

Heat-treating steel castings improves their physical properties. Metallurgists who wish to make certain that a given product will meet certain physical characteristics make castings of steel containing alloying elements. The properties of low-alloy steels (such as hardness and resistance to abrasion) are controlled largely by the response of these steels to heat treatment.

High-alloy steels take advantage of some specific property of the alloy, such

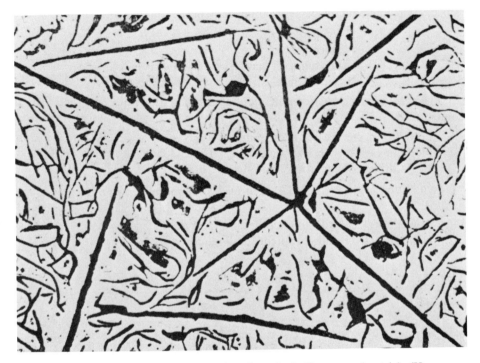

Fig. 11–10. Graphite in hypereutectic gray iron (unetched; 60 × approximately). (Hanneman and Schrader) (Courtesy A. P. Gagnebin – International Nickel Company)

as resistance to corrosion or heat, or some property other than hardenability. Examples are the outstanding wear resistance of a 14% manganese steel or the corrosion resistance of an 18% chromium stainless steel.

Cast steel has many desirable properties, but its casting requirements tax the ability of the designer, the metallurgist, and the foundryman. It has high pouring temperatures (2900–3000°F) (1593–1649°C), and thus requires special consideration from the standpoint of refractories, molding sands, metal handling, and pouring and molding techniques. Steel has a high rate of solidification shrinkage, and thus produces problems of risering and design which cast iron does not. Molten steel also reacts with oxygen and other impurities. Therefore, to ensure the production of castings of a uniformly high quality, one must adopt certain special melting procedures.

Steel is made up of the same five basic elements as iron: carbon, silicon, manganese, phosphorus, and sulfur. However, the carbon content must be less than 2%. The remaining elements are usually in the following amounts:

Manganese, %	0.5–1.0
Silicon, %	0.2–0.8
Phosphorus, max%	0.05
Sulfur, max%	0.06

Plain carbon steels have commercial classifications determined by their carbon content.

1. Low-carbon steel (carbon less than 0.20%).
2. Medium-carbon steel (carbon between 0.20 and 0.50%).
3. High-carbon steel (carbon more than 0.50%).

In addition to these three classes of plain carbon steels, two other classes are defined.

1. Low-alloy steels (alloy content totaling less than 8%).
2. High-alloy steels (alloy content totaling more than 8%).

Table 11–5 The principal effects of major alloying elements in steel

Element	Percentage	Primary function
Manganese	0.25–0.40	Combines with sulfur to prevent brittleness
	>1%	Increases hardenability by lowering transformation points and causing transformations to be sluggish
Sulfur	0.08–0.15	Contributes to ease of machining metal
Nickel	2–5	Acts as a toughener
	12–20	Improves corrosion resistance
Chromium	0.5–2	Increases hardenability
	4–18	Improves corrosion resistance
Molybdenum	0.2–5	Forms stable carbides; inhibits grain growth
Vanadium	0.15	Forms stable carbides; increases strength while retaining ductility; promotes fine grain size
Boron	0.001–0.003	Acts as powerful hardening agent
Tungsten		Enhances hardness at high temperatures
Silicon	0.2–0.7	Increases strength of spring steels
	2	
	Higher percentages	Improves magnetic properties
Copper	0.1–0.4	Improves corrosion resistance
Aluminum	Small	Makes a good alloying element in nitriding steels
Titanium		Fixes carbon in inert particles
		Reduces martensitic hardness in chromium steels

Table 11–5 lists some of the elements used as alloys in steel, and the effects of each.

Manganese and silicon are usually already in steel, as residuals from the refining process. When steel is cooled from an elevated temperature, its hardness is increased by manganese and silicon.

However, when sulfur is present in amounts of more than 0.06%, it has a harmful effect on both ductility and toughness.

Excess phosphorus causes steel to become brittle at low temperatures, which can cause cold shortness in castings.

Manganese, silicon, and phosphorus are soluble in iron, and thus do not reveal themselves in the microstructure.

Iron that is free from carbon changes its crystallographic structure twice while it is cooling from an elevated temperature to room temperature. The first change takes place at 2555°F (1401°C). At this point, the iron changes from a body-centered-cubic structure to a face-centered-cubic structure (recall Fig. 6–2). At 1670°F (910°C), the iron changes back from a face-centred-cubic structure to a body-centered-cubic structure. This phenomenon is known as an *allotropic transformation*. It is because of this feature that it is possible to control the properties of steel, by (1) controlling the carbon content, and (2) utilizing the process of heat-treatment.

Figure 11–11 is a portion of the iron-carbon diagram. The allotropic changes can be seen on the left edge of the diagram. (Recall that there was a schematic representation of the iron-carbon diagram in Fig. 6–13, and that Fig. 6–21 showed the structure of a mild steel with 0.2% carbon.)

Ferrite (which is iron) accepts only slight amounts of carbon at room temperature. Carbon which dissolves in excess of this slight amount appears in the form of iron carbide or cementite. Cementite is associated with ferrite in a lamellar structure called pearlite (recall Fig. 6–16). When there is 0.83% carbon, the entire microstructure of the steel is a mixture of ferrite and carbide or pearlite. When the carbon is 0 to 0.83% ferrite and pearlite appear in separate patches. Ferrite is softer and more ductile than pearlite, and thus affects the properties of cast steel.

When the carbon content is greater than 0.83%, cementite is present with the pearlite. Thus in these higher-carbon steels, cementite appears in large areas around pearlite areas.

The strength and hardening characteristics of a cast steel are directly related to the percentage of carbon in it.

Steel castings can be found abundantly in the automotive and earth-moving industries, in electrical machinery, and in all kinds of manufacturing. Cast steels have played an important role in the development of rockets, jet aircraft, and nuclear power. These industries have a steadily increasing need for metals which have good strength characteristics and a high degree of resistance to corrosion and creep at high temperatures. Steel castings have kept pace with their needs.

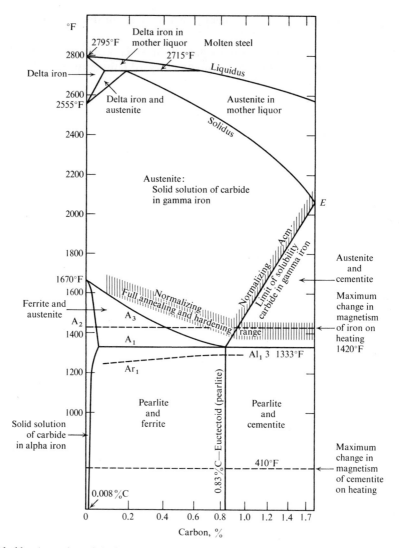

Fig. 11–11. A portion of the iron–iron carbide diagram. (Courtesy Steel Founders' Society of America.)

BIBLIOGRAPHY

1. R. W. Heine, P. C. Rosenthal and C. R. Loper, *Principles of Metal Casting,* second edition, McGraw-Hill, New York, 1967

2. A. P. Gagnebin, *The Fundamentals of Iron and Steel Castings,* The International Nickel Company, New York, 1957

3. Carl A. Samans, *Engineering Metals and their Alloys*, Macmillan, New York, 1957

4. *Malleable Iron Castings*, Malleable Founders' Society, Cleveland, Ohio, 1960

5. *The Gray Iron Casting Handbook*, Gray Iron Founders' Society, Cleveland, Ohio, 1958

6. *Steel Castings Handbook*, third edition, Steel Founders' Society, Cleveland, Ohio, 1960

CHAPTER 12

MELTING OF METALS

We may well define the melting of metal as the changing of the metal from a solid to a liquid state by the application of heat. Since the casting process depends on the ability of an alloy of metals to flow while it is in its molten state and fill a cavity, the person in charge of the casting must maintain tight control during the melting process.

There are several different ways to melt metals; each involves different equipment. The method of melting and the equipment used depend on the temperature required, the economics and scale of the installation and operation, the quantity of molten alloy needed, and the makeup of the alloy.

The types of furnaces used in the melting of alloys include the following.

1. *Crucible furnace*

   a) Stationary, fuel-fired
   b) Tilting, fuel-fired

2. *Reverberatory furnace*, fuel-fired and stationary or tilt

3. *Electric-arc furnace*

   a) Indirect-arc
   b) Direct-arc
   c) Resistance

4. *Induction furnace*

   a) Low-frequency
   b) High-frequency
   c) Channel
   d) Coreless

5. *Open-hearth furnace*

6. *Cupola furnace*

   Each of these has certain advantages and disadvantages.

CRUCIBLE FURNACES

Crucible furnaces, the simplest and oldest vehicles for the melting of metals, may be of the stationary lift-out type (Fig. 12.1a) or of the tilting type (Fig. 12.1b). A

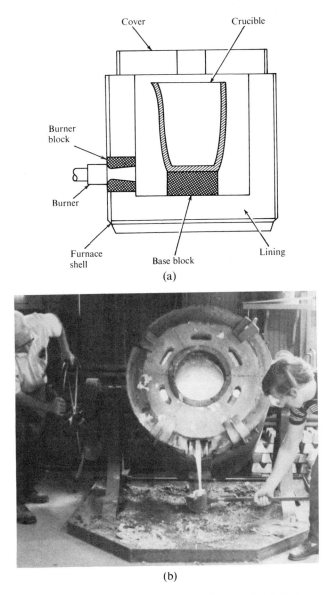

Fig. 12–1. (a) Cross section of a stationary crucible furnace. (b) A tilting crucible furnace. (Photograph courtesy Pensylvania State University.)

crucible furnace usually consists of a firebrick-lined steel shell with a movable cover. A crucible made of either a clay-graphite or a clay-silicon-carbide mixture is used within the furnace. The capacity of the furnace is of course limited by the size of the crucible. (See Table 12.1.) Lift-out crucibles make possible the greatest amount of flexibility with respect to the range of alloys that can be melted. One can control the quality of the metal by, for one thing, using different crucibles for melting each different alloy; this prevents the contamination of one alloy by another.

Table 12–1 Standard sizes of graphite crucibles

No.	Height outside, in.	Diameter top out, in.	Diameter bilge out, in.	Diameter bottom out, in.	Approx. capacity, lb, water full*	Approx. working capacity, lb, red brass
0000	$2\frac{15}{16}$	$2\frac{3}{8}$	$2\frac{3}{8}$	$1\frac{3}{4}$	0.25	1.19
1	$3\frac{3}{8}$	$3\frac{1}{4}$	$3\frac{1}{4}$	$2\frac{1}{4}$	0.50	2.96
2	$4\frac{1}{2}$	$3\frac{3}{4}$	$3\frac{11}{16}$	$2\frac{7}{8}$	0.75	4.74
3	$5\frac{3}{8}$	$4\frac{1}{4}$	$4\frac{1}{8}$	3	1.0	8.5
4	$5\frac{3}{4}$	$4\frac{5}{8}$	$4\frac{9}{16}$	$3\frac{1}{4}$	1.50	10.07
6	$6\frac{1}{2}$	$5\frac{1}{4}$	$5\frac{1}{4}$	$3\frac{7}{8}$	2.25	15.41
8	$7\frac{1}{8}$	$5\frac{7}{8}$	$5\frac{7}{8}$	$4\frac{1}{4}$	3.0	20.74
10	$8\frac{1}{16}$	$6\frac{1}{16}$	$6\frac{9}{16}$	$4\frac{15}{16}$	4.81	36.0
12	$8\frac{1}{2}$	$6\frac{3}{8}$	$6\frac{7}{8}$	$5\frac{1}{16}$	5.0	42.0
14	$8\frac{7}{8}$	$6\frac{11}{16}$	$7\frac{3}{16}$	$5\frac{1}{4}$	5.75	48.0
16	$9\frac{1}{4}$	$6\frac{15}{16}$	$7\frac{1}{2}$	$5\frac{1}{2}$	7.18	53.0
18	$9\frac{13}{16}$	$7\frac{5}{16}$	$7\frac{15}{16}$	$5\frac{13}{16}$	8.6	64.0
20	$10\frac{5}{16}$	$7\frac{11}{16}$	$8\frac{3}{8}$	$6\frac{1}{8}$	10.0	74.0
25	$10\frac{15}{16}$	$8\frac{3}{16}$	$8\frac{7}{8}$	$6\frac{1}{2}$	12.0	89.0
30	$11\frac{1}{2}$	$8\frac{5}{8}$	$9\frac{5}{16}$	$6\frac{13}{16}$	14.0	104.0
35	12	9	$9\frac{3}{4}$	$7\frac{1}{8}$	16.0	119.0
40	$12\frac{1}{2}$	$9\frac{3}{8}$	$10\frac{1}{8}$	$7\frac{7}{16}$	18.0	134.0
45	$13\frac{3}{16}$	$9\frac{7}{8}$	$10\frac{11}{16}$	$7\frac{13}{16}$	21.0	157.0
50	$13\frac{3}{4}$	$10\frac{1}{4}$	$11\frac{1}{8}$	$8\frac{1}{8}$	24.0	179.0
60	$14\frac{7}{16}$	$10\frac{13}{16}$	$11\frac{11}{16}$	$8\frac{9}{16}$	28.0	209.0
70	$15\frac{1}{16}$	$11\frac{1}{4}$	$12\frac{3}{16}$	$8\frac{15}{16}$	32.0	239.0
80	$15\frac{5}{8}$	$11\frac{11}{16}$	$12\frac{11}{16}$	$9\frac{1}{4}$	36.0	269.0
90	$16\frac{3}{16}$	$12\frac{1}{8}$	$13\frac{1}{8}$	$9\frac{9}{16}$	40.0	298.0
100	$16\frac{11}{16}$	$12\frac{1}{2}$	$13\frac{1}{2}$	$9\frac{7}{8}$	44.0	328.0
125	$17\frac{3}{8}$	13	$14\frac{1}{16}$	$10\frac{5}{16}$	50.0	373.0
150	$18\frac{3}{8}$	$13\frac{3}{4}$	$14\frac{7}{8}$	$10\frac{7}{8}$	60.0	468.0
175	$19\frac{1}{4}$	$14\frac{3}{8}$	$15\frac{9}{16}$	$11\frac{3}{8}$	70.0	523.0
200	20	15	$16\frac{1}{4}$	$11\frac{7}{8}$	80.0	597.0
225	$20\frac{3}{4}$	$15\frac{1}{2}$	$16\frac{13}{16}$	$12\frac{5}{16}$	90.0	672.0
250	$21\frac{3}{8}$	16	$17\frac{5}{16}$	$12\frac{11}{16}$	100.0	747.0
275	22	$16\frac{7}{16}$	$17\frac{13}{16}$	13	110.0	822.0
300	$22\frac{1}{2}$	$16\frac{7}{8}$	$18\frac{1}{4}$	$13\frac{3}{8}$	120.0	896.0
400	$24\frac{5}{16}$	$18\frac{3}{16}$	$19\frac{11}{16}$	$14\frac{7}{16}$	160.0	1195.0

*To find capacity in other alloys, multiply water capacity by specific gravity of metal.

Most crucible furnaces are fired by either gas or oil because these fuels are generally available and are easily controlled. In some cases, electrical energy, coal, or coke is used as the source of heat. Sometimes crucible melts are also made by setting the crucible inside a high-frequency electrical coil; it then becomes a lift-coil type of furnace.

Alloys which melt at a relatively low temperature, such as aluminum and copper-base alloys, are generally melted in a crucible furnace. Ferrous metals are not generally melted in this type of furnace, however, except in the case of induction-heated furnaces.

A crucible furnace should be charged first with clean scrap (gates, risers, and sprues) and then ingots. One should take care to avoid wedging larger pieces against the wall of the crucible, since this might cause a rupture of the wall as the metal expands during heating. Scrap should be cleaned (that is, any molding sand adhering to it should be removed by sand-blasting or wire-brushing) and metal ingots should be preheated by placing them on the cover of the furnace for a short time before opening the cover and injecting the new metal into the molten metal that is already in the crucible. When you preheat the metal, you eliminate the danger that accompanies the placing of the cold metal into the liquid metal. When cold metal is added to a hot furnace, moisture condenses on the surface of the cold metal. This moisture can turn to steam and cause an explosion, which blows molten metal out of the crucible and furnace and may cause persons near the furnace to suffer serious burns. The first pieces of metal placed in a cold crucible need not be preheated, of course, but all metal added subsequently during the melting operation must be preheated.

The linings of furnaces should be maintained in good condition, so that the fuel will be completely burned. This lessens the possibility that hydrogen will be present and be absorbed by the molten metal. Water vapor—present in the burning fuel, the refractory material, the metal, or the products of combustion—is another contaminant to be guarded against. It is therefore necessary that the crucible metal and the furnace be dry, and that during melting, the atmosphere in the furnace be slightly oxidizing.

REVERBERATORY OR AIR FURNACES

The *air furnace* is a modified open-hearth design in which the burning fuel is injected without regenerative action, from only one end (Fig. 12.2). No regenerative action takes place because it is the blown-in burning fuel that heats the furnace and not the air heated by contact with brick previously heated by outgoing air or gases. The fuel, bituminous lump coal, is pulverized by high-speed rotary crushers and blown into the furnaces as fine dust. Oil may be used as a fuel also.

In furnaces which are capable of handling only small batches (Fig. 12.3), the fuel may be natural gas or a combination of oil and natural gas. The products of combustion are drawn across the metal charge and out the draft stack.

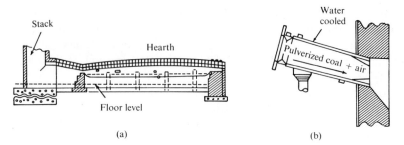

Fig. 12–2. (a) Sectional view of a typical air furnace. (b) Water-cooled streamline burner. (Courtesy Whiting Corporation, Harvey, Illinois.)

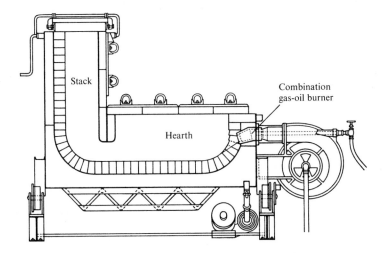

Fig. 12–3. Sectional view of a small gas- and/or oil-fired air furnace for melting brass, bronze, gray iron, and iron alloy. All the metal is charged at one time into the top of the stack, where it is preheated by the gases passing through it until it melts. The melted metal flows down into the flat hearth, where the temperature is quickly raised to the desired point. (Reda Pump Company, Bartlesville, Oklahoma.)

The air furnace is chiefly used for melting iron to be used in the production of white iron for *malleableizing*—that is, the metal is structurally changed through the heat-treating process—or of other high-grade cast irons. The melting capacity of such furnaces may range from approximately 15 tons to as large as 75 tons per batch. A typical operating cyle—melting and then tapping the batch—may take about seven hours. Air furnaces are also used to receive and hold metal that has been melted in the cupola. In malleable iron practice, this is known as *duplexing*.

ELECTRIC-ARC FURNACES

Electric-arc furnaces are melting furnaces which obtain heat generated from an electric arc within the furnace. When an arc is maintained between two electrodes, the furnace is known as an *indirect-arc furnace* (Fig. 12.4). When an arc is maintained between an electrode and the metal being melted, it is known as a *direct-arc furnace*. The direct-arc furnace generally has three electrodes, each connected to one of the leads of a three-phase power source.

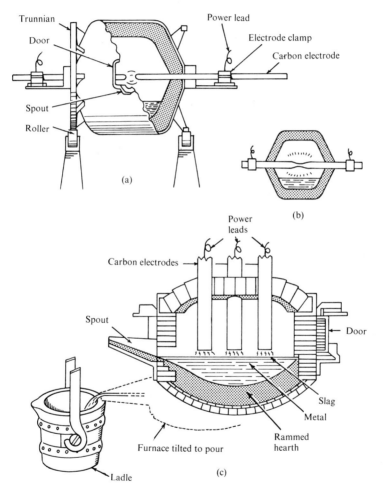

Fig. 12–4. Electric-arc furnaces. (a) Indirect-arc furnace, "rocking arc"; (b) resistance unit; (c) direct-arc furnace (three-phase). (Courtesy George E. Schmidt, Jr., M.I.T.)

Arc furnaces have either a basic or an acid lining. The chemical composition of the metal thus depends on the development of the correct slag.

The electrodes used in melting steel are made of carbon, and use fairly low voltages and high current flow. Large current-carrying bus bars and lead-in cables from the transformer are necessary. When high-temperature metals such as molybdenum are to be melted, the electrodes may be made of the same metal as the melt. This is known as *consumable arc melting*. Alloys with low melting temperatures are not melted in arc furnaces, but tool steels and other high-quality alloy steels *are* generally produced in them.

INDUCTION FURNACES

In an *induction furnace*, the metal is heated by means of either a high-frequency or a low-frequency electromagnetic field. High-frequency furnaces of the lift-coil type are limited by the size of the crucible. Rammed-refractory tilt-type high-frequency furnaces are similarly limited.

In the melting furnace illustrated in Fig. 12–5 the crucible has a coil of copper tubing wrapped around it. The coils conduct a high-frequency (1000–30,000 cps) alternating current at a high power level. An alternating magnetic field is created, which in turn induces a high alternating current in the surface of the metal inside the furnace. The heat generated by this current is conducted into the center and the entire mass of metal is quickly melted. Water is circulated through the copper coils as a coolant to prevent the furnace from breaking down due to overheating. Since the metal is melted rapidly and there are no combustion products present, there is very little loss from oxidation, and the metal is much cleaner. However, it goes without saying that clean scrap or charge material must be used.

In a low-frequency furnace which has twin coils for melting, pouring capacities may range from 200 to 5000 pounds of aluminum. A low-frequency furnace, when started, requires a *heel* of molten metal, 19–100 pounds of metal at the bottom of the crucible. Thus they are emptied only when they have to be cleaned.

A *coreless induction furnace* is a cylindrical furnace with a primary coil backed by a magnetic flux. Yokes surround the lower portion of a refractory-lined melting hearth. When an ordinary 60-cycle current is applied, heat is generated, by induction, in the metal charge. Line-frequency induction furnaces can be used for the melting, holding, superheating, or duplexing of all ferrous and nonferrous metals.

Some coreless furnaces operate at 60, 180, 540, and 960 cycles, and others at frequencies of 3000, 10,000 and even as high as 100,000 cycles. Furnaces with the higher frequencies readily melt metal from a cold charge. But those with lower frequencies require that a molten heel of metal be kept on hand so that solid metal can be added to the liquid metal and be melted.

The 60-cycle line-frequency coreless induction cold-charge unit has developed into an efficient and economical furnace. Stirring action is greater in the 60-cycle unit than in the higher-frequency units, and such a furnace quickly melts down additional charges when a liquid heel is present.

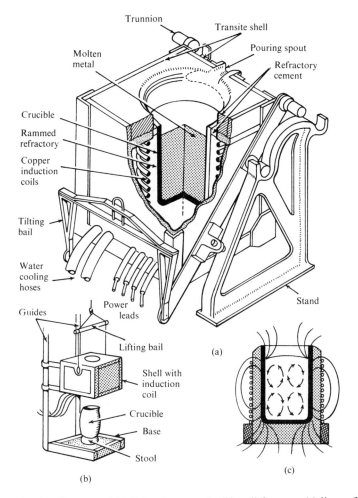

Fig. 12–5. Induction furnaces. (a) Tilting furnace, (b) lift coil furnace, (c) lines of magnetic force and stirring action on the molten-metal bath. (Courtesy George E. Schmidt, Jr., M.I.T.)

The *channel-type* or *cold induction furnace* is an efficient melting unit. The induction coil is essentially immersed within the bath of molten metal (Fig. 12–6). Liquid metal in the melting channels—plus electromagnetic forces—circulate the metal in the hearth. This circulation provides for uniform temperature and composition of the alloy throughout the bath. Since heat is generated in the metal itself, metal losses are reduced to a minimum. Thus the operation of this furnace is relatively cool and clean.

The *cored furnace* is a production-type furnace that lends itself well to melting, holding, and duplexing, and is used when relatively long periods of continuous operation are desirable. However, it requires a molten heel of metal to start, and thus is not suitable for intermittent operations.

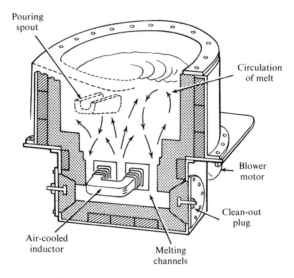

Fig. 12–6. Cored induction furnace.

OPEN-HEARTH FURNACES

The *open-hearth furnace* is the most commonly used in the production of steel (see Fig. 12–7). It is a squat, rectangular brick structure that has a shallow, elliptical hearth to hold the metal charge. Figure 12–8 shows a cross-sectional view of such a furnace.

The size of an open-hearth furnace is based on the weight of metal that it is expected to produce per batch. In steel foundries, the capacities range from 10 to 125 tons per batch. In steel mills, open-hearth furnaces sometimes are capable of producing as much as 300 tons per batch.

Two types of melting practices—acid melting and basic melting—refer to the chemical reactions between the refractory lining and the slag. In the acid melting process, the furnace is lined with a silica brick. However, with this method, it is harder to control the amount of sulfur and phosphorus in the molten steel. A high-grade material is needed to make up the charge of metal to be melted. In the basic melting process, the furnace has a magnesia-brick lining, which makes possible the removal of sulfur and phosphorus. Thus one can use a lower grade of pig iron and steel scrap. Lime is used with the charge, as a fluxing or slag-producing agent.

An open-hearth furnace is heated by means of gas or oil; in some operations powdered coal is used. The fuel is burned in two combustion chambers, one at each end of the furnace. One burner is used for about 15 or 20 minutes, then the other. The air used for combustion should be preheated to help maintain the high temperature required in the open-hearth process. This is accomplished by passing the gases over brick checkerworks at each end of the furnace. The checkers are heated alternately by passing hot gases through them. The preheated checkers then help increase the temperature of the incoming hot gases on the next cycle. This is known as *regenerative heating* of the fuel gases.

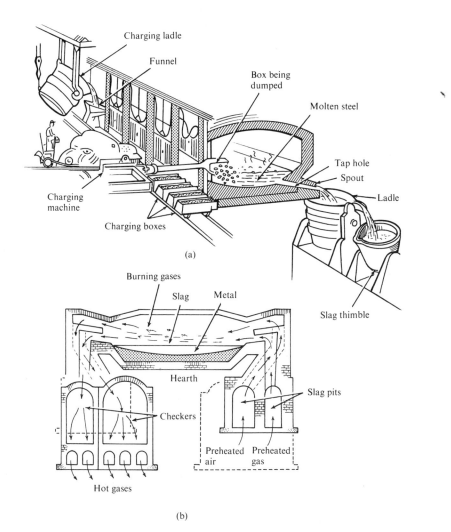

Fig. 12–7. Open-hearth furnace. (a) Pictorial views, (b) longitudinal cross section. (Courtesy George E. Schmidt, Jr., M.I.T.)

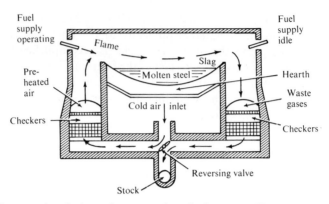

Fig. 12–8. Cross-sectional view of an open-hearth furnace. (Courtesy Steel Founders' Society of America.)

The hot gases entering the hearth heat and melt the metal, and at the same time also heat the lining. The heated lining reflects the heat back into the furnace, which also helps melt the metal. The open-hearth furnace is thus reverberatory in its action. The metal bath (pool of molten metal) in an open-hearth furnace is shallow so that there is a maximum surface area for a given volume of metal, and thus heat is transferred readily. This feature also facilitates the reaction between metal and slag, and thus increases the refining of the metal.

SIDE-BLOW CONVERTER

Some foundries produce steel in a *side-blow converter* process (Fig. 12–9), in which an air blast is used to oxidize and remove the carbon, silicon, and manganese contained in the molten cast iron. The iron is obtained from the cupola and given a treatment of soda ash so that it forms a fluid slag that reacts rapidly with the sulfur in the cupola metal.

After the converter has received the charge of molten iron, the air blast is turned on and the refining process begins. The characteristics of the flame coming from the mouth of the converter vary as the process continues. Measuring the flame by means of a photoelectric-cell device gives a good indication of the condition of the steel. When the refining is finished, the metal will have about 0.05% C, 0.02 to 0.05% Si, and 0.01 to 0.03% Mn.

At this point the steel may be prepared for pouring by deoxidizing and recarburizing to adjust the chemical composition.

When liquid metal is transferred to an electric holding furnace, further adjustments of its composition can be made; thus greater flexibility is made possible.

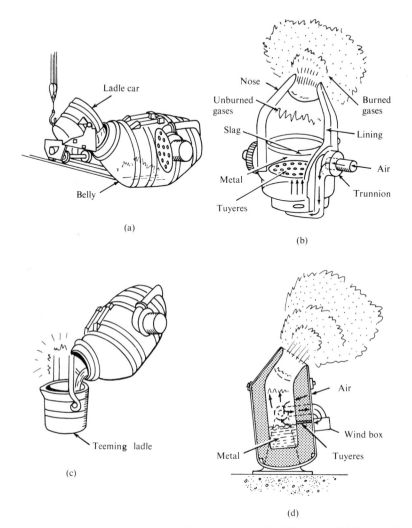

Fig. 12–9. Converters. Bessemer bottom-blow converter. (a) Charging, (b) blowing, (c) tapping, (d) Tropenas side-blow converter. (Courtesy George E. Schmidt, Jr., M.I.T.)

BASIC OXYGEN FURNACE

The *basic oxygen furnace*, which can make steel faster than any other device, is replacing the open-hearth furnace. When molten pig iron is available, operating costs of this furnace are low, and it can produce high-quality steels that are low in phosphorus, sulfur, nitrogen, and hydrogen.

The basic oxygen furnace is a top-blown modification of the pneumatic converter, using industrially pure oxygen instead of air (Fig. 12–10). Top-blowing

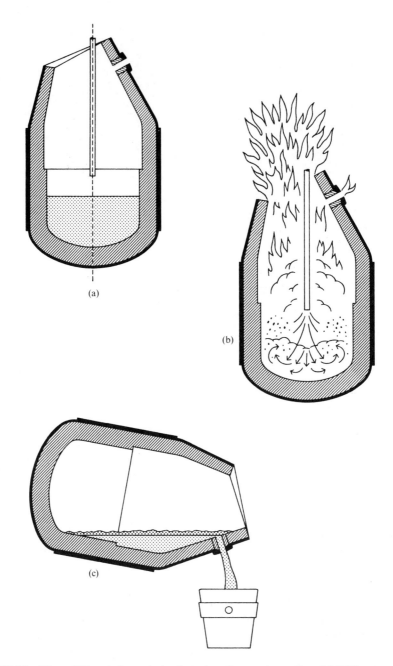

Fig. 12–10. Three different phases in basic oxygen furnace operation. (a) With oxygen lance inserted, (b) during blowing operation, and (c) during tapping operation. (From F. E. Van Voris.)

with a water-cooled lance is used because at the higher temperatures reached it is not possible to blow oxygen through bottom tuyeres (see the Glossary of Terms) without excessive refractory damage. The lining, naturally, is basic.

The total time from charge to tap runs between 40 minutes and one hour. Large furnaces can produce steel at a rate of 100 tons per hour.

After the furnace has been charged with scrap and hot metal, it is positioned vertically and the lance is lowered into the center of the furnace. Oxygen is blown in at about 100 ft^3/min per ton of charge. The reaction between the slag and the metal takes place rapidly because of the high oxygen input ratio, the high temperature, and the constant mixing of slag and metal.

At such high temperatures, iron oxide is produced in the reaction zone of the furnace. The loss in yield is small (1.0%), but the fume contains such finely divided particulate matter that it causes serious pollution of the air. However, when relatively economical equipment for washing and cleaning the gas is incorporated into the design of the furnace, the fume is absent. Therefore the basic oxygen furnace can be made to meet even the most stringent air cleanliness codes.

Advantages of the process are: a great degree of dephosphorization and partial removal of sulfur; a steel output which has low residual contents of oxygen, hydrogen, and nitrogen; low capital outlay; flexibility; and the speed with which steel is produced.

THE CUPOLA

In the production of iron, the *cupola* is used to remelt and recover scrap iron created during the forming of gray iron, malleable iron, and ductile iron. The final mix is adjusted to the desired composition by additions of pig iron and alloys.

Cupolas are operated to produce commercially iron which has a carbon content ranging from 2.40 to 3.80%, with melting rates ranging from 1 to 30 tons per hour. Diameters of commercial cupolas vary from 18 to 84 inches. The weight of charges varies from 140 to 5400 lb.

All cupolas use coke as the principal fuel; they now use by-product coke, beehive coke, and anthracite coal. In addition, they sometimes use other materials as supplements when carbon is absorbed during the melting process due to a deficiency in the regular coke or a reduction in the pig-iron content of the charge. Such materials are pitch coke, petroleum coke, calcined pitch coke, carbon electrodes, and graphite blocks. A few cupolas operate with gas burners located above the tuyeres to reduce the consumption of coke.

The cupola is a vertical steel shaft lined with refractory, equipped with air ports (tuyeres) located around the outer periphery of the shell near the bottom (see Fig. 12–11) and a charging door in the upper section, for introducing the raw materials. Below the tuyeres is a section called the *well*, at the bottom of which is the tap hole which allows molten metal to be withdrawn. The slag is removed from the top of the well and just below the tuyeres. In front-slagging, the metal and slag discharge through a single hole, and pour into a small well in the runner, where the slag is dammed off for removal.

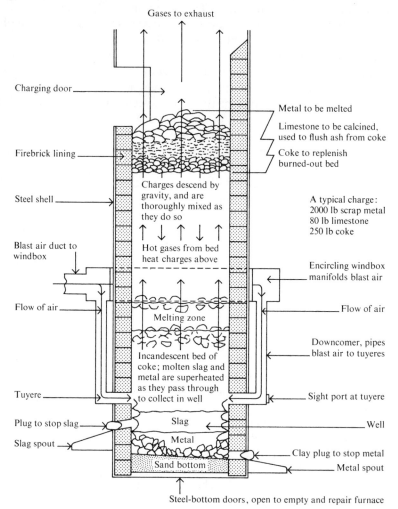

Gases to exhaust

Charging door

Metal to be melted

Limestone to be calcined, used to flush ash from coke

Firebrick lining

Coke to replenish burned-out bed

Charges descend by gravity, and are thoroughly mixed as they do so

Steel shell

A typical charge: 2000 lb scrap metal 80 lb limestone 250 lb coke

Blast air duct to windbox

Hot gases from bed heat charges above

Encircling windbox manifolds blast air

Flow of air

Melting zone

Flow of air

Downcomer, pipes blast air to tuyeres

Incandescent bed of coke; molten slag and metal are superheated as they pass through to collect in well

Tuyere

Sight port at tuyere

Plug to stop slag

Slag

Well

Slag spout

Metal

Clay plug to stop metal

Sand bottom

Metal spout

Steel-bottom doors, open to empty and repair furnace

Fig. 12–11. Schematic diagram of a typical cupola.

Initially heat for the process is supplied by a bed of coke, on the top of which are placed alternative layers of iron, coke, and limestone. Air from a blower introduced through the tuyeres burns the coke, and the hot gases ascend through the upper charges of metal. As the coke in the bed is consumed, the heat released melts the metal. Then the fresh coke between the charges reaches the bed and the process continues. The process is one of countercurrent flow, with the heat and gases rising upward and out the top of the stack, while the metal descends and is withdrawn from the bottom of the well. The process is thermally efficient not only because of this countercurrent flow but also because the materials are always in intimate contact with each other.

The simplicity of the process is the reason for both its success and the many difficulties it presents. Due to the variables and the variations, the chemistry of the process is as complex as the mechanics are simple.

Dimensions and corresponding melting rates of cupolas are shown in Table 12–2. The melting rate is a direct function of the diameter of the cupola.

Melting capacities of cupolas

Diam., in.	Area, in²	Lb/hr at 10 lb/in²/hr	Tons/hr
36	1018	10,180	5.1
48	1810	18,100	9.0
60	2827	28,270	14.1
72	4072	40,720	20.3

These rates of melting are generally increased by increasing the rates of flow of air and/or increasing the air temperature by blast preheating, or by decreasing the rates of flow of air or by increasing the amount of coke. Cupola capacity may be indicated by the inside diameter at the top of the coke bed, where the iron is melted, or by the rated output for a cupola of that inside diameter. The melting rate in turn may be determined not only by the diameter but also by the ratio of coke to iron after the bed is formed. This ratio is expressed as a whole number denoting the pounds of iron melted per pound of coke used.

Depending on the particular conditions of combustion, the amount of air by weight will be equal to or greater than the amount of metal charged. Each ton of coke requires eight tons of air. It is the amount of oxygen in the air which effects element and reaction activity in the combustion of the coke to the carbon gases.

A cupola melt of any duration should be fluxed with limestone. The undesirable materials to be removed are coke ash, rust, dirt on scrap, and sand on returns. Flux is also an aid in reducing sulfur.

Fluxing furnishes a nonsoluble (in iron) liquid as a medium to absorb extraneous material. Fluxing also lowers the melting point of the components of the refractory so that it is liquid enough at existing temperatures to separate readily from the iron; thus it can be removed. This nonmetallic covering on the molten metal, caused by the use of a flux, is called *slag*. If slag is concentrated in the same area each time during melting, excessive burnout of the furnace lining can result. Bridging of the coke and slag across the cupola can occur if too little flux material is used. Improper fluxing can also result from poor patching of the side walls of the cupola after each melt.

Table 12–2 Cupola dimensions (courtesy Whiting Corporation, Harvey, Illinois)

Cupola size	Shell diam.	Min. thickness of lower lining*	Diam. inside lining	Area inside lining, in²	Melting rate, tons/hr, with iron/coke (after bed) ratios of				Coke and iron charges, lb					Limestone, lb	Air through tuyeres, ft³/min	Suggested blower selection†	
										Iron							
					6	8	10	12	Coke	6/1	8/1	10/1	12/1			ft³/min	discharge pressure oz
0	27″	4½″	18″	254	¼	1			20	120	160			4	570	640	8
1	32″	4½″	23″	415	1	1½			35	210	280			7	940	1040	16
2	36″	4½″	27″	572	1¾	2¼			45	270	360			9	1290	1430	20
2½	41″	7″	27″	572	1¾	2¼			45	270	360			9	1290	1430	20
3	46″	7″	32″	804	2½	3¼	4		65	390	520	650		13	1810	2000	24
3½	51″	7″	37″	1075	3¼	4¼	5¼		85	510	680	850		17	2420	2700	24
4	56″	7″	42″	1385	4	5½	7		110	660	880	1110		22	3100	3450	24
5	63″	9″	45″	1590	4½	6¼	8		130	780	1040	1300		26	3600	4000	28
6	66″	9″	48″	1809	5½	7¼	9	10¾	145	870	1160	1450	1740	29	4100	4500	32
7	72″	9″	54″	2290	7	9¼	11½	13¾	185	1100	1480	1850	2220	37	5200	5750	32
8	78″	9″	60″	2827	9	11¼	14	17	225	1350	1800	2250	2700	45	6400	7100	32
9	84″	9″	66″	3421	10½	13¾	17	20½	275	1650	2200	2750	3300	55	7700	8600	36
9½	90″	9″	72″	4072	12¼	16¼	20¼	24½	325	1950	2600	3250	3900	65	9200	10200	36
10	96″	9″	78″	4778	15	19	23¾	28¾	385	2300	3080	3850	4600	77	10700	11900	36
11	102″	12″	78″	4778	15	19	23¾	28¾	385	2300	3080	3850	4600	77	10700	11900	36
12	108″	12″	84″	5542	17	22¼	27¾	33¼	445	2670	3560	4450	5400	89	12500	13900	36

*For long heats, use heavier linings.

†Additional pressure capacity may be required when auxiliary equipment is added to the blast system, or when piping is long or complicated.

The carbon equivalent is generally calculated as: C.E. = % TC + $\frac{1}{3}$(%Si + %P).

The materials used as charges for the cupola are pig iron, return iron (runners, gates, and risers), and purchased iron and steel scrap. One must be selective as to the kind of purchased scrap used. Burned iron, hard or chilled iron, nonferrous materials, or stove plate may lead to the production of poor iron. For good chemical balancing of the iron, use cleaned, graded scrap.

During the melting process, losses and pickup of various elements occur. Carbon can go to a maximum of 4.3% and then balance out, since loss of carbon due to oxidation equals pickup of carbon from coke. Approximately 10% of silicon is lost through oxidation, and the silicon level must be maintained by the addition of pig iron, ferro-silicon, or silicon briquette. Manganese losses—due to oxidation and reaction with sulfur—approximate 20%. Sulfur, because it is an undesirable element in iron, should be held to less than 0.10%. There is no loss of sulfur during melting, but there often is a pickup of sulfur from the metal charged and from the coke. The pickup from coke is approximately 4%. Phosphorus is usually held to a maximum of 1%; there is little loss or gain of phosphorus during melting.

Table 12–3 shows the loss or gain of elements as a percentage of the weight of each element charged. Table 12–4 gives the approximate analysis of cast iron and steel scrap.

Table 12–3 Approximate loss or gain of elements in acid melting

Element	% loss	% gain
Silicon, in pig iron and scrap		
Si 3% in metal charge	7–12	—
Si 2% in metal charge	7–12	—
Si 1% in metal charge	7–12	—
Si 0.50% in metal charge	7–12	—
Lump ferrosilicon	10–15	—
Manganese in pig iron and scrap	10–20	—
Lump ferromanganese	15–25	—
Spiegeleisen	15–25	—
Chromium, lump ferro	10–20	—
Nickel, shot or ingot	2–5	—
Copper, shot or $\frac{3}{16}$ and thicker scrap	2–5	—
Alloy in briquettes	5–10	—
Sulfur	—	40–60

All materials charged into the cupola must be weighed accurately. This is essential for several reasons:

1. One must properly balance the materials used in order to arrive at the correct chemical and metallurgical composition.

2. It is necessary to keep a proper inventory of materials used and an exact control of the cost of operation.

3. One needs to keep accurate, dependable records of the various mixes used in order to duplicate results and to correct difficulties.

Table 12–4 Approximate analysis of cast iron and steel scrap*

Type of scrap		% Si	% total carbon	% Mn	% P max.	% S
No. 1 Mach. Cast	lt.	2.00–2.50	3.40–3.60	0.50	0.30	0.14
and	med.	1.50–2.00	3.25–3.50	0.60	0.15	0.14
No. 1 cast	hvy.	1.25–1.50	3.00–3.20	0.75	0.15	0.14
Automotive cylinders		2.00	3.20	0.65	0.08	0.08
Truck and tractor cylinders		2.25	3.40	0.75	0.12	0.14
Agricultural scrap		1.80–2.50	3.25–3.60	0.50–0.75	0.25–0.50	0.14
Pipe (water)		1.25–1.85	3.40–3.65	0.35–0.45	0.30–0.70	0.10–0.12
Radiator scrap		2.20–2.50	3.40–3.70	0.50–0.60	0.35–0.55	0.12
Brake shoes		0.90–1.10	3.10–3.30	0.20–0.40	0.30–0.40	0.18–0.25
RR car wheel chilled Fe		0.50–0.60	3.25–3.50	0.50–0.60	0.25–0.35	0.14
Malleable scrap		1.10–1.50	2.20–2.50	0.30–0.50	0.10–0.15	0.08–0.12
Rails		0.20–0.30	0.60–0.80	0.70–0.80	0.05	0.05
Auto steel		0.10–0.30	0.10–0.50	0.50–0.80	0.05	0.05
Structural steel		0.10–0.30	0.10–0.20	0.50–0.80	0.05	0.05

*Alloying elements (such as Cu, Ni, Mo) may be present and should be determined by analysis.

Table 12–5 shows how to calculate a chemical balance for a charge in a simple arithmetic manner.* Each component is calculated as follows: Pig iron contains $2.40 \times 0.30\%$ silicon, and since pig iron represents 30% of the mix, it will contribute 2.40×0.30 or 0.72% silicon to the final mix. During the melting operation, there will be a loss of 10% of silicon and 15% of manganese. At the same time there is a pickup of sulfur, generally about 0.03% for coke with a normal sulfur content and being used in the usual quantities. If the sulfur in the coke or the quantity of coke is greater than normal, the sulfur pickup increases proportionately. Carbon pickup cannot be calculated, because it depends on the state of oxidation within the cupola.

* Tables 12–3, 12–4, and 12–5 are all reprinted from *The Cupola and Its Operation*, published by the American Foundrymen's Society, 1965.

Table 12-5 Daily mixture calculation sheet

Mixture no.   Heat no.   Date

Cupola no.

Material charged	%	lb/chg.	Carbon %	lb	% in mix*	Silicon %	lb	% in mix*	Manganese %	lb	mix %*	Sulfur %	lb	% in mix*	Phosphorus %	lb	mix %*
Pig iron, mall.	20	400	4.30	17.20	0.86	0.90	3.60	0.18	0.55	2.20	0.110	0.03	0.12	0.006	0.15	0.60	0.030
Pig iron, foundry	4	80	2.50	2.00	0.100	7.5	6.00	0.30	0.65	0.52	0.026	0.05	0.040	0.002	0.10	0.08	0.004
Silvery piglets																	
50% Fe Si lump																	
Purchased scrap	26	520	3.30	17.16	0.858	1.90	9.88	0.494	0.50	2.60	0.130	0.12	0.624	0.031	0.35	1.82	0.091
Cast-iron briquets																	
Steel briquets																	
Steel scrap	20	400	0.20	0.80	0.040	0.20	0.80	0.040	0.60	2.40	0.120	0.04	0.160	0.008	0.03	0.12	0.006
Returns	30	600	3.40	20.40	1.020	1.85	11.10	0.555	0.50	3.00	0.150	0.12	0.720	0.036	0.19	1.14	0.057
Other: Mn, Cr, Ni, Cu, Mo, etc.			Late Fe Si addition 13.1 lb of 75% $\frac{3}{8}$ × 12M			9.81	0.49										
Total	100	2000		57.56	2.878		41.19	2.05		10.72	0.536		1.664	0.083		3.76	0.188
Analysis charged %	(Totals divided by weights)				2.88			2.05			0.54			0.08			0.188
Percent melting gain or loss					+0.44			−0.20			−0.05			+0.04			+0.010
Estimate analysis (%) at cupola spout					3.32			1.85			0.49			0.12			0.198

*Percent of element (C, Si, Mn, S, P) based on 2000-lb charge.

BIBLIOGRAPHY

1. R. W. Heine, P. C. Rosenthal, and C. Loper, *Principles of Metal Casting*, McGraw-Hill, New York, 1968

2. H. F. Taylor, M. C. Flemings, and J. Wulff, *Foundry Engineering*, John Wiley, New York, 1959

3. American Foundrymen's Society, *The Cupola and Its Operation*, third edition, Des Plaines, Ill., 1968

4. F. E. Van Coris, and J. Crane, "Basic Oxygen Steelmaking," *AFS Trans.* **68**, page 756, 1960

5. R. Rinesh, H. Neudecker, and J. Eible, "Basic Oxygen LD Process for Foundries," *J. Steel Castings Res.*, April 1963

CHAPTER 13

CLEANING AND INSPECTION

To the foundryman, the term *cleaning operations* generally refers to the removal of gates and risers from the casting, and the removal of any mold or core sand that might be adhering to it. Any other foreign material, such as wires or chills, are likewise removed at this time. There may also be a grinding operation to remove sprue or risering irregularities and fins that might have cropped up during the casting operation. After the casting has been cleaned, there is a visual inspection of it for minor defects of salvageable castings, and then an inspection of the finished casting. This could encompass nondestructive testing. In steel foundries, weld repairing of surface defects is handled by the cleaning department.

REMOVAL OF SAND FROM CASTINGS

After shakeout of the casting, any sand that adheres to it is removed either by vibrating the casting on a cleaning table, wire-brushing it, or by blast cleaning. *Blast cleaning* is the quickest and most uniform method of removing sand and scale, and it improves the as-cast surface finish. There are three types of blast cleaning: air, water, and mechanical blasting. When air is used as the carrying medium, it is made to propel sand, shot, grit, or glass beads. This operation must be confined to an enclosed room or cabinet with an adequate dust-control system. If air blast cleaning can conveniently be carried out with the gates and risers remaining attached, they will be cleaned at the same time as the casting, and thus when they are remelted they will be free of sand and other contamination. This naturally improves the condition of the metal.

In water blasting, water is used as the carrying medium to propel sand against the casting. The big advantage of water blasting is that it eliminates the dust problem.

OTHER MEANS OF SURFACE CLEANING

Perhaps the most efficient of all methods of cleaning castings used in production foundries today is the mechanical or airless-blast method. Both small and large castings are cleaned in this manner. Small castings are caused to rotate or tumble under blasts of grit or shot. Large castings are placed on stationary or rotating tables inside a cabinet, and particles of shot or grit are thrown at them by means of centrifugal force derived from a rapidly rotating wheel, as illustrated in Fig. 13–1. All surfaces of a casting are exposed to this abrasive action. Sometimes castings are suspended from moving conveyor hooks which carry them into a room in which shot or grit is mechanically thrown at all their exposed surfaces. The shot or grit used may be made of white iron, malleableized iron, or steel. Shot blasting has a peening effect on the casting surface (which becomes flattened by the hammering of the steel shot thrown with force against it) and can actually cause a surface-metal flow. Grit has a much harsher effect on the surface of the casting than shot does, sometimes removing small particles of metal from the surface. Shot blasting produces a shiny surface, while grit dulls the surface.

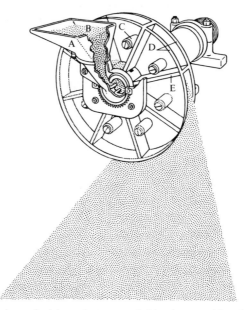

Fig. 13–1. Phantom view of airless shot- or grit-blasting machine head (wheelabrator). (Courtesy The American Wheelabrator and Equipment Co.)

When a nonferrous casting is treated by shot blasting, the force propelling the shot must be reduced or the harsh action of the shot will severely damage the surface of the casting. Copper or aluminum castings are too soft to withstand the action of grit blasting. For treating castings made of nonferrous metals, soft steel, copper, or bronze, glass beads are quite effective. Excessive amounts of sand or metal particles left over from previous cleanings retard the effectiveness of the shot used in airless blasting, and must be removed from the shot during the recycle action.

Wire brushing, through mechanical means, is often sufficient to clean the surface sand from aluminum, brass, or bronze alloy castings, and leads to the production of a shiny surface.

REMOVAL OF GATES, RISERS, AND FINS

The cutting or trimming of gates, risers, fins, or flashing may take place either before or after the cleaning operation. When physically practical, one ought to perform this cutting operation after the casting has been cleaned, in order for the metal which will subsequently be remelted, to be clean metal.

The *tumbling barrel* is another device for cleaning sand from castings. A rotating iron drum or barrel is filled with castings and then small star-shaped pieces of iron are added. The castings are tumbled with the mill stars and the two abrade each other. However, the burnishing action on the surfaces of the castings may cause the corners of the castings to be rounded if they are tumbled excessively. Wet tumbling, involving the addition of water treated with a caustic, may be used to keep down dust. When copper-base castings are tumbled about with steel or iron balls, sand, or pumice stone in combination with water and detergents, their surface finish is improved considerably.

The removal of gates, risers, and other protuberances which are not part of the casting is accomplished by *mechanical cutoff machines* employing band saws, abrasive cutoff wheels, and metal shears. Shears are best for the more ductile materials. Large risers and gates on steel castings are readily removed by the cutting torch and oxygen lance. The size of the torch nozzle and the oxygen pressure necessary to keep the cutting reaction moving are factors that depend on the thickness of the metal to be cut.

Powder cutting is a process by which large risers cast from oxidation-resistant alloys can be cut. Preheated iron powder is introduced into an oxygen stream; this burning iron then attacks the metal riser by a process of fluxing and oxidation. In this manner, one can rapidly remove gates from castings made of alloys which are highly resistant to oxidation.

The trimming of fins, flashing, and other similar appendages is often accomplished by chipping by hand tools or with pneumatic chipping hammers (Fig. 13–2). Portable grinding tools, stationary-stand grinders, and swing-frame grinders are used for rough grindings to remove excess metal. The use of either aluminum oxide or silicon carbide types of abrasive has improved the removal of metal and the grinding of castings. Varied sizes of abrasive used at certain defined speeds, can produce a finish that is either smooth or coarse, as desired. A wide space between grains of abrasive material is often used for rapid removal of metal when the surface is to be machined later.

In many cases in which the defects in castings are not too serious, the castings can be salvaged by means of welding. The economics of a given situation—size, shape, casting alloy, and cost of a new or replacement casting—help determine the salvage criteria. Some castings that leak when they are given a pressure test may be salvaged by a sealing process which takes place in an autoclave. Sodium silicate or various types of resin serve as impregnants which fill small cavities and seal the castings. This should not be attempted unless it is acceptable to the user of the casting.

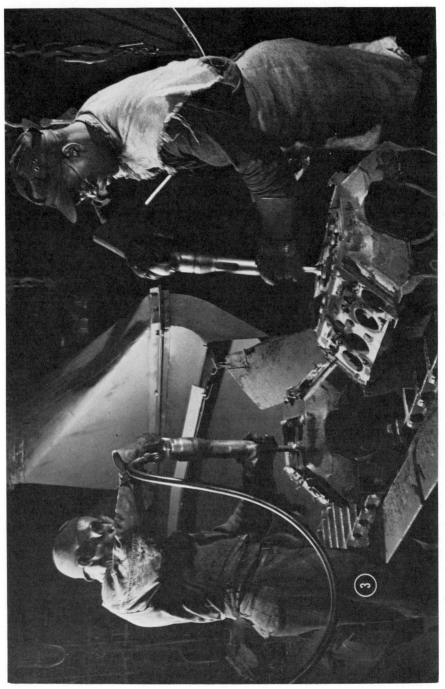

Fig. 13–2. Operators use pneumatic hammers to punch the intake valve portholes in cylinder block castings.

NONDESTRUCTIVE TESTING

In recent years, nondestructive testing of materials, components, and assemblies has become a widely used practice, as mass-production methods have also become widely used. As the standards of quality control continue to improve, customers in the future, instead of simply requiring that the castings delivered to them be free of defects, will undoubtedly increasingly demand nondestructive testing of the castings before delivery.

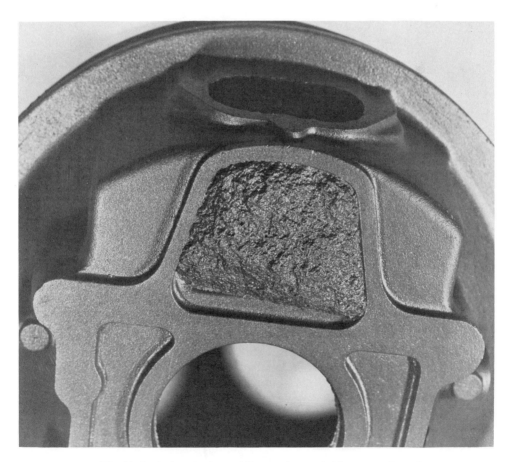

Fig. 13–3. A drop in the molding sand caused this excess of metal.

The term *nondestructive testing* is used to describe all methods which make possible the testing or inspection of a material without impairing its future usefulness. From an industrial viewpoint, the purpose of nondestructive testing is to determine whether a material or part will satisfactorily perform its intended purpose.

The number of variables present in the manufacture of castings, especially sand castings, accounts for the rather high rate of loss due to defective castings. Most defects can be prevented if the person in charge is really careful and brings as many of the variables as possible under control. In order to reduce the incidence of a given defect, and possibly eliminate it, one must first assemble all the facts,

Fig. 13–4. Back draft on the cracking strip weakens the sand pocket and permits a drop.

then use them to determine the nature of the defect, the possible causes, and eventually the remedy for it. Trial castings are used to test the study and prove the results of the investigation.

Some of the more frequently encountered defects of castings are listed below, and illustrated in Figs. 13–3 through 13–15. (This list is not intended to be complete.)

1. *Blowholes,* including open blows. Both appear as smooth-walled voids or cavities, either on the surface or within a casting. Both result from gas or steam created during the casting process.

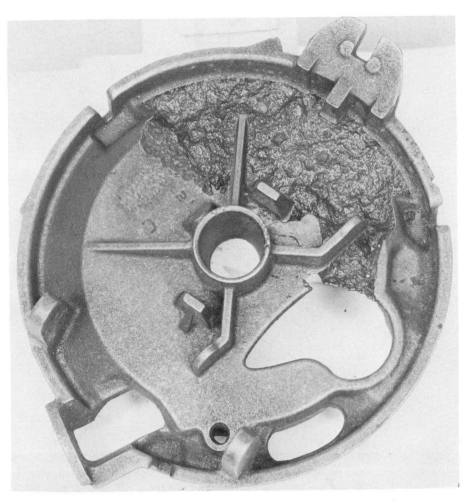

Fig. 13–5. Inadequate green strength in the molding sand creates a drop.

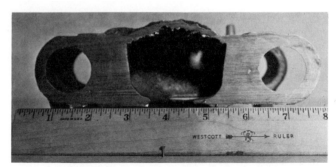

Fig. 13–6. A low shear strength permits this pocket of sand to shear to the point that metal flows under the drop.

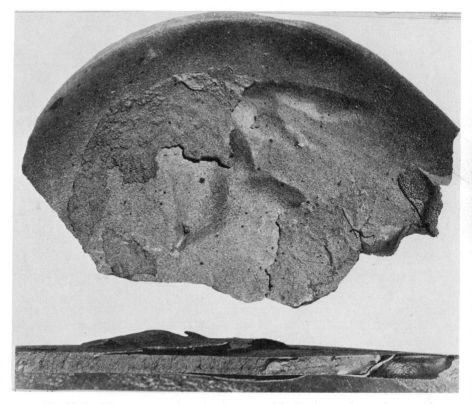

Fig. 13–7. Blister, on a sanitary casting, caused by hard ramming and wet sand.

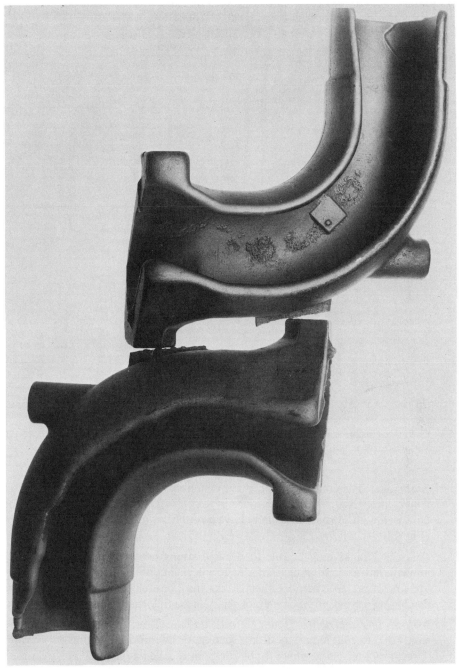

Fig. 13–8. Misrun caused by too-small gates which froze before casting filled.

Fig. 13–9. Expansion scab caused by slow pouring which permits excessive heating of the cope surface before metal contact.

2. *Shrinkage.* The volumetric shrinkage which takes place when metal solidifies, if not compensated for, may cause a void. Voids due to shrinkage can be confused with blows, but usually have a rougher wall and an irregular shape. (Fig. 13–14).

3. *Cracks.* Cracks may be due to the low strength of some alloys beginning immediately after solidification, due to internal cooling stresses. Such cracks are known as *hot tears.* Other cracks may be due to improper handling, machining, or design. Some cracks may be so tiny that one cannot see them.

4. *Misruns and cold shuts.* These defects are caused by improper filling of the cavity due to cold metal, too thin a section, or by improper gating or pouring. A section of a casting which is not filled out is called a *misrun*; two streams of metal not completely fused create a *cold shut* (Fig. 13–8).

5. *Inclusions.* Undesirable foreign material trapped in or on a casting is known as an inclusion. Common inclusions include oxides, slag, loose sand, and gas in the melt.

6. *Pinhole porosity.* This is the result of the gas inclusions referred to above. Most molten metal tends to pick up gases, such as oxygen and hydrogen, which often end up as very small holes throughout the casting.

7. *Rattails and buckles.* These are generally associated with sand expansion. Both appear on the surface of the casting. A rattail is a thin irregular line of metal raised above the surface; a buckle is a thicker raised spot.

8. *Other defects* include core shift, mold shift, swell, runout, drop, and other special defects. (Figs. 13–12, 13–13, 13–15).

In order to spot defects, one must carefully inspect castings and also parts produced by other manufacturing processes. The methods of inspection may be classified as destructive or nondestructive, depending on whether or not the part is destroyed or used up during the inspection process.

The simplest and most common inspection is *visual inspection.* Every person who handles a part during manufacturing is essentially an inspector with respect to surface defects. Visual inspection does not, of course, tell anything about the internal conditions of the casting, nor does it reveal minute surface cracks.

Fig. 13–10. Scab near the ingate caused by nonuniform heating of the mold surface.

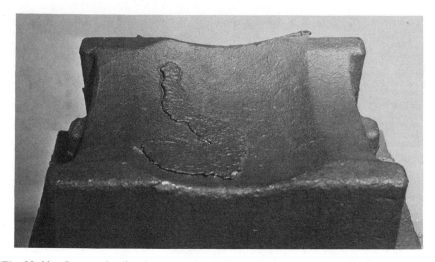

Fig. 13–11. Cope scab related to excessive moisture in the sand aggravated by hard ram.

Fig. 13–12. A typical mold shift is shown where a cope and drag do not match.

Fig. 13–13. A typical core raise resulted in an apparent misrun over the core.

Closely related to simple visual inspection is *dimensional inspection*. Not all parts have to conform to size specifications, but when they do, measuring them with special gages as well as common measuring devices provides the checks. If a part is to be machined, the casting, especially the first few castings to come out of a given mold, should be checked to make certain that pattern and core boxes are correct.

More exacting inspection can be accomplished with special methods. These procedures can be classified in four general categories.

1. *Penetrant inspection.* Penetrant testing methods are used to inspect surfaces for minute cracks which are difficult to detect visually. A penetrant dye applied to the surface may indicate the presence of a crack. After one removes the excess dye and applies a special developer, the defects are clearly indicated, as shown in Fig. 13–16. Nonmetallic substances may be inspected with equally accurate results by this method which utilizes a liquid with an excellent wetting and penetrating ability. The penetrating liquid contains either a material which will fluoresce under black light or a dye that can be visually detected after a developer is added.

If a fluorescent penetrant is used, defects show up as glowing yellow-green dots or lines against a dark background. In dye penetrants, defects are indicated as red dots or lines against a white background. Typical defects are illustrated in Fig. 13–17.

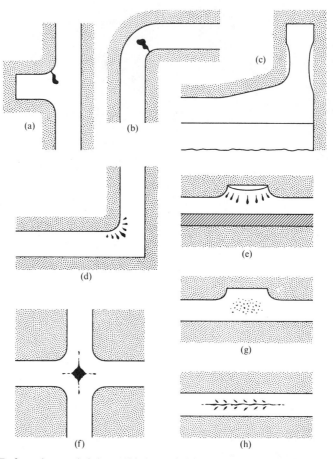

Fig. 13–14. Defects due to shrinkage. (a) An unfed hot spot. (b) Internal unsoundness with wall puncture due to inadequate feeding. (c) Dishing of surface due to inadequate feeding. (d) and (e) Wormholes at an internal angle or on a cope surface. (f) Hot spot causing internal shrinkage cavity. (g) Pinholes caused by imperfect feeding. (h) Center-line shrinkage. (From Taylor, Fleming, and Wolff, *Foundry Engineering*, Wiley, New York, 1959.)

Interpreting the characteristic patterns of different types of flaws is very important. A crack or cold shut is indicated by a line of penetrant. Dots of penetrant indicate pits or porosity. A series of dots indicates a tight crack, cold shut, or partially welded lap. Accurate interpretation can be learned only by experience.

Broadly speaking, dye-penetrant techniques identify voids which are open to the surface. They do not identify internal porosity, shrinkage, or other defects *not* open to the surface. People use the penetrant processes mainly to detect flaws in nonferrous or nonmagnetic ferrous alloys. The penetrant system is very useful when one is dealing with nonmetallics materials such as ceramics and plastics. It can be used routinely for the examination of defects in castings.

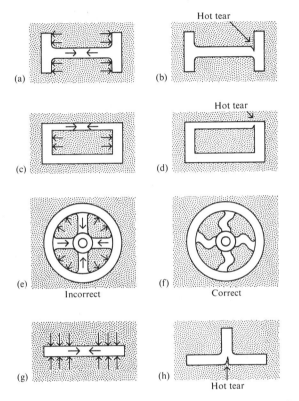

Fig. 13–15. Defects due to a mold's preventing the casting from contracting freely. (a) and (b) Stresses developed in the casting by resistance of the mold. (c) and (d) Thinner parts of the casting resist normal contraction, while heavier sections cool more slowly. (e) and (f) Spokes that are curved rather than straight reduce warping or tearing. (g) Uniform section cools without tearing. (h) Nonuniform section or joined sections provide potential site of tearing. (From Taylor *et al.*, *Foundry Engineering*, Wiley, New York, 1959.)

2. *Magnetic-particle inspection.* The detection of small cracks on or near the surface of ferromagnetic objects is often best accomplished by magnetic-particle techniques. In a magnetized object, if a crack or void interrupts the magnetic field, the magnetic field is distorted.

In a material which has an induced magnetic field, defects such as blowholes, cracks, and inclusions produce a distortion. The path of the magnetic flux is distorted because the defects have magnetic properties different from those of the

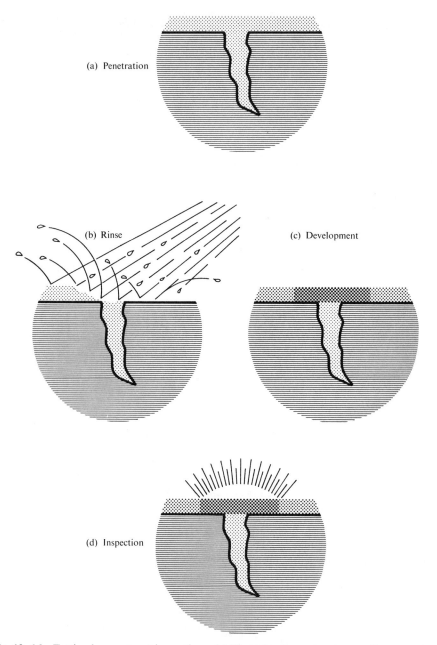

(a) Penetration

(b) Rinse

(c) Development

(d) Inspection

Fig. 13–16. Testing by penetrant inspection. (a) Fluorescent penetrant on surface seeps into crack. (b) Watery spray removes penetrant from surface, but not from cracks and pores. (c) Developer acts like a blotter to draw penetrant out of crack. (d) Black light causes penetrant to glow in dark.

surrounding material. The magnetic flux spreads out in order to detour around the interruption, and flux lines extend outside the metal. Small magnetic particles show the path of the flux line, and hence the shape of the crack or void. The magnetic field makes an angle of 90 degrees with the suspected crack. Figure 13–18 shows the leakage field produced by a surface and a subsurface flaw.

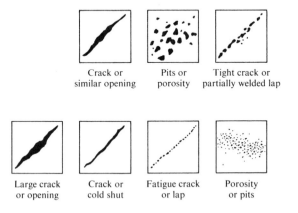

Crack or similar opening	Pits or porosity	Tight crack or partially welded lap

Large crack or opening	Crack or cold shut	Fatigue crack or lap	Porosity or pits

Fig. 13–17. Indications of defects. (Courtesy Turco Products, Inc.)

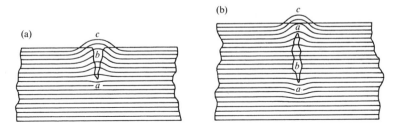

Fig. 13–18. (a) Leakage field due to surface flaw. (b) Leakage field due to subsurface flaw.

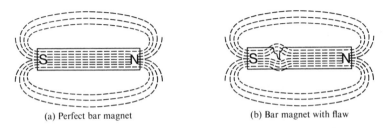

(a) Perfect bar magnet   (b) Bar magnet with flaw

Fig. 13–19. The disturbance due to a flaw. Weak north and south poles are set up across the crack and the magnetic particles tend to bridge the gap. A defect is indicated by a black line of Fe_3O_4 whose width may be 200 times that of the original crack, and is therefore easily visible.

This method is quick and cheap to operate, very sensitive, and can be applied to nearly all steel and iron castings, except austenitic materials. The basic principle is that, although magnetic flux lines are caused to flow in a metal object, the local disturbance at a flaw is detected by using a magnetic powder (either dry or as a suspension in a liquid). Iron filings may be used, but ferric oxide is preferred.

Figure 13–19 shows the disturbance due to a flaw. The magnetic particles have been treated with a fluorescent material, so that the interruption must be detected by a black-light source.

3. *Ultrasonic testing.* Internal defects like those that can be detected with radiography may also be detected with sound. Ultrasonic testing utilizes high-frequency sound waves produced by a quartz crystal. A sound pulse is introduced into the metal and the time intervals between transmission of the outgoing and reception of the incoming signals are measured with a cathode ray oscilloscope. The pattern on the oscilloscope can be analyzed to give both the location and the size of the internal defect (Fig. 13–20).

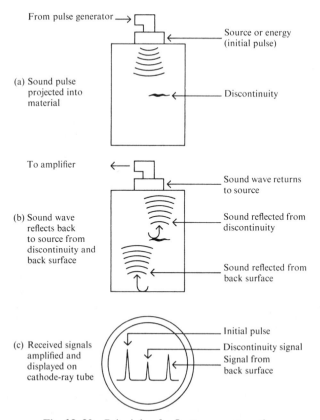

From pulse generator →

Source or energy (initial pulse)

(a) Sound pulse projected into material

Discontinuity

To amplifier

Sound wave returns to source

(b) Sound wave reflects back to source from discontinuity and back surface

Sound reflected from discontinuity

Sound reflected from back surface

(c) Received signals amplified and displayed on cathode-ray tube

Initial pulse

Discontinuity signal

Signal from back surface

Fig. 13–20. Principle of reflectoscope operation.

4. *Radiography.* Certain short-wavelength rays can penetrate dense materials and react with an emulsion on special film, which makes possible this method of inspection (Fig. 13–21). Very dense material withstands penetration and causes the emulsion to be affected to a lesser degree. Such areas on the film have a light appearance, whereas less-dense materials allow more penetration, and the film appears darker. Any hole or crack within the casting is thus revealed as a dark area.

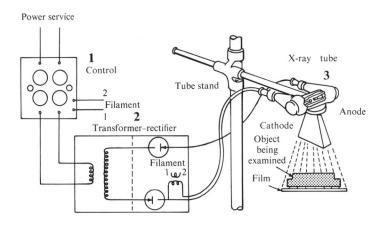

Fig. 13–21. Schematic diagram of a specimen set up for making an x-ray of a step casting.

Radiography includes both x-ray and gamma-ray radiation, the difference being in the source of the rays; x-rays are the result of a sudden change in the velocity of a moving electric charge. Powerful equipment is needed to produce x-rays; equipment capable of penetrating thick sections of iron or steel require voltages of up to 15,000,000.

Gamma rays are emitted by radium in sealed containers, as well as by pile-produced isotopes such as cobalt-60. These sources emit rays in all directions; hence, to avoid injury to personnel, their use must be carefully controlled.

Comparatively heavy sections of castings used in power plants employing atomic energy require 100% radiographic examination. For optimum examination of wall junctions in a reasonable amount of time, engineers use *betatron radiography*, which provides a high-energy source of radiography. A 3000-lb casting whose heaviest section is 6 inches thick requires a 2-hr exposure, while a 16,000-lb casting whose heaviest section is 14 inches thick requires a 24-hr exposure.

The operation of the betatron is fast, sensitive, and comparatively simple. The quality of a casting can be determined quickly even in the early stages of production.

Cobalt-60 sources, which produce gamma rays, require a much longer period of time and are occasionally limited in their effectiveness when it comes to heavy cross sections of castings.

EDDY CURRENTS

A coil which carries alternating current causes an *eddy current* to flow in any nearby metal. The eddy current may react on the coil to produce substantial changes in its reactance and resistance, depending on the proximity and conductivity of the metal. If the path of the eddy current is distorted and lengthened by even a small crack or defect in the metal, the reactance and resistance of the coil are again modified.

The eddy current is confined to a *surface layer*, whose depth depends inversely on the frequency of the current and on the conductivity and permeability of the metal. Thus the eddy-current method can be applied only to the detection of cracks at or near the surface.

Figure 13–22 illustrates the application of the method. Induced current is used to find circumferential defects. Testing time is reduced to less than half the time required by the magnetic-particle method, since there is no necessity for the time-consuming process of clamping and making a current pass twice through positions 90 degrees apart. Care must be used in setting up the casting for testing, however, as the magnetizing coil must be kept at a constant distance from the piece of metal to be tested.

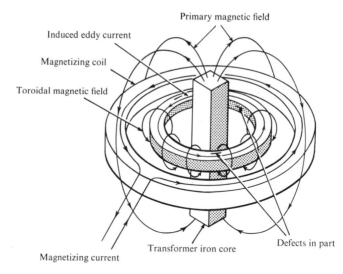

Fig. 13–22. Schematic illustration of the application of eddy currents.

BIBLIOGRAPHY

1. C. A. Hogarth and J. Blitz, *Techniques of Nondestructive Testing*, Butterworths, London, 1960
2. W. J. McGonnagle, *Nondestructive Testing*, McGraw-Hill, New York, 1961
3. R. F. Hanstock, *The Nondestructive Testing of Metals*, The Institute of Metals, London, 1951

CHAPTER 14

CASTING DESIGN CONSIDERATIONS

A designer of metal objects, in the last analysis, is looking mainly for the least-expensive way to make a given component which adequately meets service requirements. There was a time when castings were used because there was no other way, regardless of cost, to attain the results desired. Today, there are relatively few parts that cannot be made successfully in several ways, and castings are used simply because they are the least expensive and because they provide the properties required.

The skills of both designers and metallurgists are needed for the proper evaluation of the service performance of a metal part. Serviceability of any metal part is a balance between applied stress (design) and the metal's resistance to the stress (metallurgy). It has been said that the designer is responsible for the applied stress and that the metallurgist is responsible for the metal's resistance to that stress. Today the designer must be familiar with some metallurgical fundamentals, as they relate to castings, and the metallurgist must be familiar with some design fundamentals.

One important factor often overlooked by the designer is the varying production methods that might be applicable to producing the part he has in mind. A designer, for example, might obtain greater economy and better quality by employing one process rather than another, and he should know the practicability and economics of both.

PATTERN DESIGN CONSIDERATIONS

Often the *quantity* of parts to be made is of prime significance, particularly from a cost standpoint. Some methods are better suited to quantity production, while other methods are better suited to the making of relatively few castings. If the part is to be a casting, the designer should produce a design such that the pattern can be adapted to standard molding methods. He should give adequate consideration to the proper placing of gates, risers, and chills to make possible a sound casting. Configuration and sizes of sections must be taken into consideration to prevent the development of stress and hot-tearing or cracking of the casting. To obtain sound castings, the designer should try to achieve directional solidification.

CAST WELD

When large castings are involved, it is often desirable to make castings of component parts, to be welded together later. Whenever possible, in the interests of simplicity of design and ease of production of the mold and casting, unwieldy projections in the form of hubs, bosses, and feet should be eliminated. Thus a designer has to be fully aware of the requirements of good foundry practice. If he is not technically capable of making proper decisions so that he achieves good castability, then he should consult the foundry engineers involved in casting the product. When the designer and producer discuss casting problems fully, the outcome can be economically beneficial to both.

Pure metals as well as alloys of metals solidify by a process of crystallization as heat energy is dissipated from the mass. Castings solidify in an inward and upward way, toward the areas of large mass, thus following a directional pattern. The larger mass, being the last to solidify, feeds the thinner sections which freeze earlier. To ensure soundness in the larger section, one uses a riser to feed it. Thus one can design a sound casting to favor directional solidification.

All castings should be designed with the path of heat flow in mind. The path of heat flow is important because this factor is what determines the pattern and mode of solidification of the casting, and hence its strength, soundness, and other features.

STRESS PLANES AND WEAKNESS

In a rectangular casting, weak places may develop at the sharp corners of the casting; this is caused by the crystals forming and growing as dendrites in columnar form. To prevent such weak places, one should use fillets which are of a proper size and which have equal cross-sectional areas.

Stresses often arise from conditions due entirely to flaws in the design. Thin sections, for example, cool much more quickly than adjacent heavier sections. As a result, these thin sections may develop high stresses, and, to relieve the stresses, may tear at the junction. Warpage may also occur due to uneven cooling of uneven sections.

At elevated temperatures, many cast metals exhibit reduced strength and ductility. Brittleness evident in hot solidified metal is known as *hot shortness*. Hot shortness, solidification contraction because of poor design, and internal voids caused by improper feeding combine to cause internal stresses that may develop into hot tears or cold cracks.

The maximum size of the cast section is often restricted by the fluidity of the metal in its liquid phase and the changes in that fluidity associated with superheat in the metal. The composition of the metal, the surface tension in the flowing metal, and the superheat all affect the fluidity of any given metal. The term *fluidity of a metal*, as used in the foundry, means the degree to which the molten metal is capable of filling a mold cavity.

COOLING RESULTS

From a metallurgical point of view, the rate at which castings cool during solidification in a mold affects the soundness of the castings. For example, gray iron castings that have been rapidly cooled exhibit chill areas due to lack of graphitization in those areas. The rate of solidification is (1) directly proportional to the rate of transfer of heat through the mold walls, (2) directly proportional to the area of the casting surfaces, and (3) inversely proportional to the mass of the casting.

Poor design can cause many flaws in castings; hot spots, for instance. And internal shrinkage cavities, as well as surface shrinkage, are usually the result of a concentrated mass existing in the thickest part of the casting, which may make it necessary for the mass to be fed by a riser to counteract volumetric contraction during solidification. Whenever possible, the design should be changed to eliminate the condition.

Abrupt changes in thickness of sections produce areas of uneven solidification; these areas often react to contraction stress by tearing. Even when the contraction is restrained, the stresses that occur can cause tearing of the casting when it is hot and thus weak. If sections are too thin or too long in relation to the fluidity characteristics of the metal, misruns and cold shuts develop; this produces an incomplete casting.

The fact of the matter is, then, that poor design is responsible for many common defects in castings, defects often attributed to laxity or mistakes on the part of the foundryman.

DESIGN RULES

Some specific rules for satisfactory design are the following.

1. Strive for uniform section size.
2. Design the casting so that there is directional solidification, by tapering sections as liberally as possible toward risers.
3. Avoid situations in which there are thin sections between heavy sections and risers.
4. Avoid designs which involve large flat surfaces.
5. Avoid multiple partings, and also irregular parting and loose pieces.
6. Try to prevent the occurrence of isolated hot spots which are difficult to feed.
7. Keep plates in tension and ribs in compression according to performance requirements.
8. Make the sections as thin as possible in view of the flowing qualities of the metal.
9. Make the sections of a size that is adequate for proper gating and risering.
10. Do not compromise the rules.

The following sketches illustrate casting design principles.
For a summary of basic design rules, see Figs. 14–1 through 14–9.

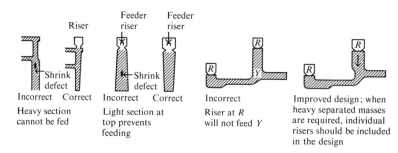

Fig. 14–1.   Design for sound castings.

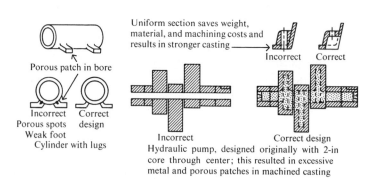

Fig. 14–2.   Design sections as uniform in thickness as possible.

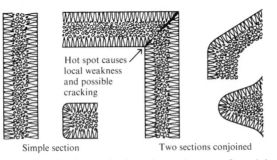

Hot spot causes
local weakness
and possible
cracking

Simple section   Two sections conjoined

Crystal structure of various casting forms shows advantages of rounded
corners; the extent of formation of columnar crystals is primarily a function
of the metal used, the rate of solidification, and the shape of the casting

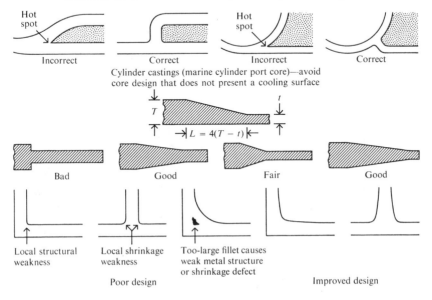

Hot
spot

Incorrect   Correct   Incorrect   Correct

Cylinder castings (marine cylinder port core)—avoid
core design that does not present a cooling surface

$L = 4(T - t)$

Bad   Good   Fair   Good

Local structural   Local shrinkage   Too-large fillet causes
weakness   weakness   weak metal structure
                                       or shrinkage defect

Poor design   Improved design

Fig. 14–3. At adjoining sections avoid sharp angles and abrupt section changes.

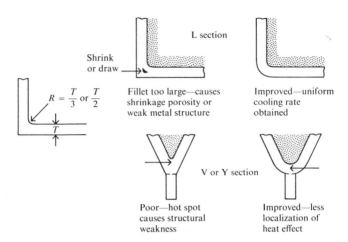

Fig. 14–4.   Fillet all sharp angles.

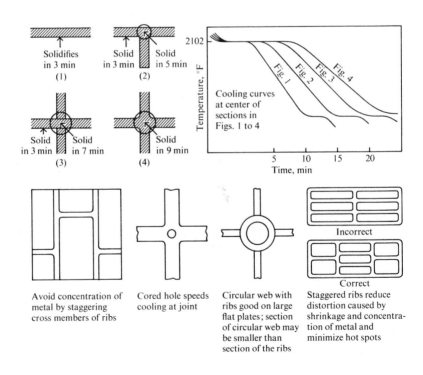

Fig. 14–5.   Bring minimum number of adjoining sections together.

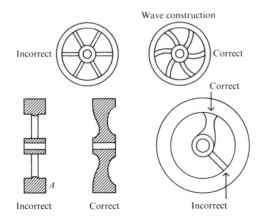

Fig. 14–6. Avoid casting strain.

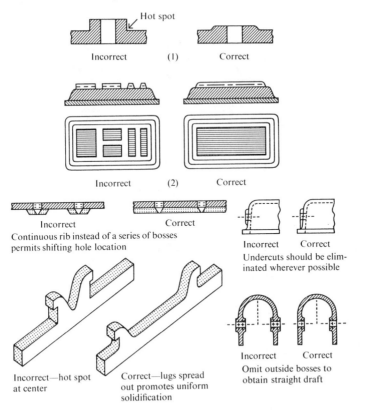

Fig. 14–7. Bosses, lugs, and pads should not be used unless absolutely necessary.

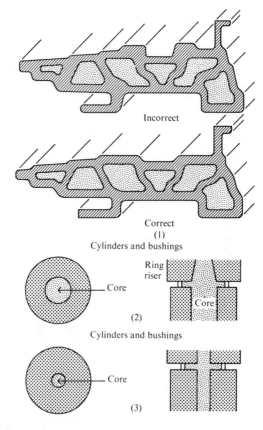

Incorrect

Correct
(1)
Cylinders and bushings

Ring
riser

Core

Core

(2)
Cylinders and bushings

Core

(3)

Fig. 14-8. Proportion dimensions of inner walls correctly.

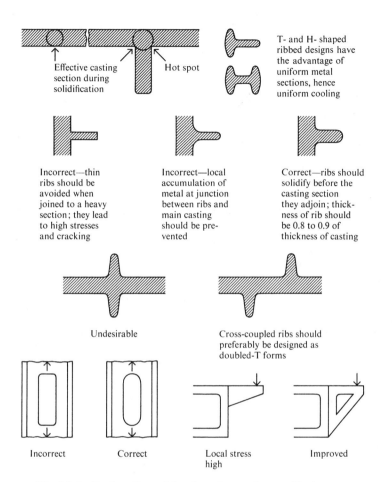

Fig. 14-9. Design ribs and brackets for maximum effectiveness.

BIBLIOGRAPHY

1. D. C. Ekey and W. P. Winter, *Introduction to Foundry Technology*, McGraw-Hill, New York, 1958

2. R. W. Heine and P. C. Rosenthal, *Principles of Metal Casting*, McGraw-Hill, New York, 1955

3. O. W. Smalley, *Fundamentals of Casting Design as Influenced by Foundry Practice*, Meehanite Metal Corporation, White Plains, N. Y., 1950

4. J. B. Caine, *Design of Ferrous Castings*, American Foundrymen's Society, Des Plaines, Ill., 1963

GLOSSARY OF TERMS

Abrasive Any substance used for abrading, such as grinding, polishing, blasting, etc. Material may be bonded to form wheels, bricks, and files, or applied to paper and cloth by means of glue. Natural abrasives include emery, corundum, garnet, sand, etc. Main manufactured abrasives are silicon carbide and aluminum oxide. Metallic shot and grit are also used as abrasives in cleaning castings.

Acicular structure A microstructure characterized by needle-shaped constituents.

Acid A chemical term to define a material which gives an acid reaction.

Acid bottom or lining (furnace) The inner bottom or lining of a melting furnace composed of materials having an acid reaction. Materials may be sand, siliceous rock, or silica brick.

Acid melting Melting in a furnace with refractory material that has an acid reaction. Material may be silica sand, siliceous rock, or silica brick.

Acid steel Steel melted in a furnace which has an acid bottom and lining, under a predominantly siliceous slag.

Additives Any material added to molding sand for reasons other than bonding or improvement of bond is considered an additive. Bonds can be of varying types: carbonaceous (seacoal, pitch, fuel oil, graphite, gilsonite); cellulose (wood flour, cereal hulls); fines (silica flour, iron oxide, fly ash); cereals (corn flour, dextrine, sugar); and chemical (boric acid, sulfur, ammonium compounds, diethylene glycol).

Aerator A device for fluffing (or decreasing the density of) and cooling sand by the admixture of air.

AFS fineness number Approximately the number of meshes per inch of a sieve which just would pass the sand sample if its grains were uniform in size. In other words, it is the average of the grains in the sand sample.

Aging A change in metal or alloy by which its structure recovers from an unstable condition produced by quenching (quench aging) or by cold working (strain aging). The change in structure consists in precipitation, often submicroscopic, and is marked by a change in physical properties. Aging which takes place slowly at room temperature may be accelerated by a slight increase in temperature.

Air belt Chamber, surrounding the cupola at the tuyeres, to equalize the volume and pressure of the blast and deliver it to the tuyeres.

Air channel A groove or hole which carries the vent from a core to the outside of a mold.

Air control equipment Any device used to regulate the volume, pressure, or weight of air.

Air dried (dry) A core or mold dried in air, without application of heat.

Air furnace A form of reverberatory furnace for melting ferrous and nonferrous metals and alloys. Flame from fuel burning at one end of the hearth passes over the bath to the stack at the opposite end of the furnace.

Air hammer Chipping hammer operated by compressed air.

Air hardening Full hardening of a metal or alloy during cooling in air or other gaseous medium from a temperature above its transformation range.

Air hoist Lifting device operated by compressed air.

Air hole Hole in a casting caused by air or gas trapped in the metal during solidification.

Air injection machine An early type of diecasting machine in which air pressure acting directly on the surface of molten metal in a closed gooseneck forces the metal into the die.

Air setting The property of some materials to take a permanent set at normal air temperature. Examples are gypsum slurry, investment molding materials, core and mold washes, etc.

Airless blast cleaning A process whereby the abrasive material is applied to the object being cleaned by centrifugal force generated by a rotating-vane-type wheel.

Allotropy Occurrence of an element in two or more modifications. For example, carbon occurs in nature as the hard crystalline diamond, soft flaky crystalline graphite, and amorphous coal.

Alloy A metallic material formed by mixing two or more chemical elements. Usually possesses properties different from those of the components.

Alloying Procedure of adding elements other than those usually comprising a metal or alloy to change its characteristics and properties.

Alloying elements Elements added to nonferrous and ferrous metals and alloys to change their characteristics and properties.

Alpha iron An allotropic (polymorphic) form of iron, which is magnetic and whose atomic structure is body-centered-cubic lattice. It is soft, ductile, of fair strength, and is capable of dissolving a few hundredths of a percent of carbon to form a solid solution.

Amorphous Having no crystalline form.

Anodizing Forming a conversion coating on a metal surface by electrolytic oxidation with the work forming the anode; most frequently applied to aluminum.

Antimony One of the elements; its chemical symbol is Sb. Its formula weight is 121.76, specific gravity 6.62, and melting point 630.5°C.

Antioch process Plaster molding process using a mixture of about 50% sand, 40% gypsum, and 8% fibrous talc mixed with water in the proportion of 100 parts by weight material with 50 parts water. After air setting (see which), the mold is placed in an autoclave and subjected to 15 psig steam pressure for 6 to 8 hours. The mold is allowed to rehydrate in air for 14 hours, and then baked at 450 to 475°F for 25 to 30 hours.

Antipiping (material) Usually refers to an insulating material placed on top of a sprue or riser that keeps the metal in liquid or semiliquid form for a long period of time and minimizes the formation of the usual conical pipe or shrink in the top of a sprue or riser.

Arbor A metal barrel, frame, or plate to support or carry part of a mold or core.

Arrestor, dust Equipment for removing dust from air.

Assembling (assembly) line   Conveyor system where molds or cores are assembled.

Atmospheric riser (Williams)   Blind riser which employs atmospheric pressure to aid feeding. Insertion of a small sand core into the riser provides a means for ingress of air into the interior of the riser, and forces the metal into the casting cavity.

Atom   The smallest particle of an element.

Austenite   A solid solution of cementite or iron carbide, Fe_3C, in iron.

Austenitic   Usual reference is to an alloy steel or iron with structure at room temperature that is normally composed essentially of austenite.

Back draft   Taper or draft which prevents removal of pattern from the mold.

Back (backing) sand   Sand between the facing sand and the flask.

Backing board   A second bottom board where molds are opened.

Baffle plate   Plate or wall in a firebox or furnace to change direction of the flame.

Bail   Connection between crane hook and ladle.

Baked core   One which has been subjected to heating or baking until it is thoroughly dry, as opposed to a green-sand core, which is used in the moist state.

Baked permeability   Property of a molded mass of sand heated at a temperature above 230°F until dry and cooled to room temperature, to permit passage of gases through it; particularly those generated during pouring of molten metal into a mold.

Baked strength   Strength of a sand mixture after it has been baked to above 230°F and cooled to room temperature.

Band, inside   A steel frame placed inside a removable flask to reinforce the sand.

Banking the cupola   Method of keeping cupola hot and ready for immediate production of hot iron after an unexpected shutdown of several hours. Procedure is to drain all molten iron and slag from the cupola, place extra coke on the top charge, and open one or two tuyeres to supply a small natural draft to keep coke combustion going.

Bar   A rib in the flask to help hold the sand.

Basic   A chemical term for a material which gives an alkaline reaction.

Basic bottom or lining (furnace)   Inner lining and bottom of a melting furnace composed of materials having a basic reaction. Materials may be crushed burnt dolomite, magnesite, magnesite brick, or basic slag.

Basic steel   Steel melted in a furnace with a basic bottom and lining under a predominantly basic slag.

Basin   A cavity on top of the cope into which metal is poured before it enters the sprue.

Batch   Amount or quantity of core or mold sand or other material prepared at one time.

Bath   Molten metal on the hearth of a furnace, in a crucible, or a ladle.

Baume   Designating or conforming to either of the scales used by the French chemist Antoine Baume in the gradation of his hydrometers for determining the specific gravity of liquids. There are two scales: one for liquids lighter than water, and the other for those heavier than water.

Bauxite   An ore of aluminum consisting of moderately pure hydrated alumina, $Al_2O_3 \cdot 2H_2O$.

Bed charge   The charge of iron placed on the coke bed in a cupola.

Bed coke   Coke placed in the cupola well to support the following iron and coke charges.

Bedding a core   Placing an irregularly shaped core on a bed of sand for drying.

Bed-in   Method of ramming the drag mold without rolling it over.

Beehive coke   Coke which is produced in hemispherical ovens about 12 ft in diameter and charged through the top to form a layer of coal 18 to 24 in. deep.   Coke is ignited and air for partial combustion is supplied over the top by doors around the bottom of the ovens.   Air burns volatile matter released by coke and during the later stages of carbonization burns some 5 to 8% of the coke.

Bellows   A device operated with both hands, to produce a current of air; some bellows are mechanically operated.

Bench   Frame support on which small molds are made.

Bench molder   Man who makes small molds on a molder's bench.

Bench rammer   A short rammer used by a bench molder.

Bentonite   A widely distributed, peculiar type of clay which is considered to be the result of devitrification and chemical alteration of the glassy particles of volcanic ash or tuff.   Used in foundry to bond sand.

Bernoulli's theorem   A theorem which states that in a stream flowing without friction, the total energy in a given amount of the fluid is the same at any point in its path of flow.

Bessemer process   Method of making steel by blowing air through molten pig or carbon-bearing iron contained in a suitable vessel which causes rapid oxidation of silicon, carbon, etc.

Binary alloy   An alloy of two metals.

Binder   Material to hold the grains of sand together in molds or cores.   May be cereal, oil, clay, resin, pitch, etc.

Binder, plastic (resin)   Synthetic resin material used to hold grains of sand together in molds or cores; may be phenol formaldehyde or urea formaldehyde thermosetting types.

Black heart   American type of malleable iron. The normal fracture shows a velvety black appearance having a mouse-gray rim.

Black lead   Graphite for facing molds and cores.

Blacking   Carbonaceous material for coating mold or core surfaces.

Blast   Air driven into the cupola or furnace for combustion of fuel.

Blast cleaning   Removal of sand or oxide scale from castings by the impinging action of sand, metal shot, or grit projected under air, water, or centrifugal pressure.

Blast furnace   Closed-top-shaft furnace for producing pig iron from iron ore.

Blast gate   Sliding plate in the cupola blast pipe to regulate the flow of air.

Blast meter   Instrument indicates the volume or pressure, or both, of air passing through the blast pipe.

Blast pressure   Pressure of air in blast pipe or wind belt of the cupola, depending on location of indicating instrument.   Usually given in ounces of water pressure.

Bleed (bleeder, bleeding)   Molten metal oozing out of casting stripped or removed from the mold before solidification.

Blended sand   Mixture of sands of different grain sizes, clay content, etc., to produce one possessing characteristics more suitable for foundry use.

Blind riser   An internal riser which does not reach to the exterior of the mold.

Blister   Defect on the surface of a casting appearing as a shallow blow with a thin film of metal over it.   In die-casting, it is a surface bubble or eruption caused by expansion of gas (usualy as a result of heating) trapped within the die-casting or beneath the plating on the die-casting.

Blocking the heat   Stopping the carbon drop in production of steel by addition of deoxidizers such as silicomanganese, spiegel, or ferrosilicon and ferromanganese.

Blow   A casting defect due to trapping of gas in molten or partially molten metal.

Blow gun   Valve and nozzle attached to a compressed air line to blow loose sand or dirt from a mold or pattern.   Also to apply wet blacking.

Blow hole   The hole or void left in a casting by trapped gas. (*See* Blow.)

Blow pipe   A small pipe or tube through which the breath is blown to remove loose sand from small molds.

Blower   Machine or device for supplying air under pressure to the melting unit.

Blower, core or mold   Machine using compressed air to inject sand into a corebox or a flask.

Blowplate   Plate on the bottom of the sand hopper on core or mold blower machines which contains holes through which the sand is blown into the corebox or flask.

Bod, bott   A piece of clay or other material to stop the flow of metal from the taphole.

Bod (bott) stick   A stick or rod on which the bod is mounted so that it may be forced into the taphole.

Body core   The main core.

Boil   Agitation of molten metal by steam or gas.

Bond   Cohesive material in sand.

Bond clay   Any clay suitable for use as a bonding material in molding sand.

Bond strength   Resistance of foundry sand to deformation.

Booking   Method of assembling or bringing together two halves of a core in a manner similar to closing a book.

Boric acid   Inhibitor used in facing sand for magnesium-base and aluminum-base alloys high in magnesium to prevent reaction with moisture in the sand.

Borings   Metal in chip form resulting from machining operations.

Boron   One of the elements.   Its chemical symbol is B and its atomic weight is 10.82.   In the form of borax and boric oxide, it is used as a flux in nonferrous metallurgy, and in the form of an alloy with other elements, as an addition to ferrous alloys.

Boron trichloride   A product used for degasification of aluminum alloys.

Bosh   Sloping of the cupola lining to form a smaller diameter just above the tuyeres.

Boss   Projection (usually of circular cross section) on a casting.

Bottom board   Board supporting the mold.

Bottom doors   Doors underneath the cupola.

Bottom pour ladle   Ladle in which metal, usually steel, flows through a nozzle in the bottom.

Bottom pour mold   Mold gated at the bottom.

Bottom sand   Layer of molding sand rammed into place on the doors at the bottom of a cupola.

Bracket   Strengthening strip or rib on a casting.

Branch core   Part of a core assembly.

Branch gate   Two or more gates leading into the casting cavity.

Brass   Copper-base alloy with zinc as the major alloying element.

Brazing   Joining metals and alloys by fusion of nonferrous alloys with melting points above 800°F, but lower than those of the materials being joined.

Breast   Area surrounding the taphole of a melting furnace.

Breeze   Coke or coal screenings.

Bridge   Material adhering to the cupola wall which slows or prevents descent of the stock charges.

Brinell hardness   Value of hardness of a metal or alloy, tested by measuring the diameter of an impression made by a ball of given diameter applied under a known load.   Values are expressed in Brinell hardness numbers.

Briquets   Compact cylindrical or other shaped blocks formed of finely divided materials by incorporation of a binder, by pressure, or both.   Materials may be ferroalloys, metal borings or chips, silicon carbide, etc.

Bronze   Copper-base alloy, with tin as the major alloying element.

Buckle   Defect on a casting surface, appearing as an identation resulting from an expansion scab.

Built-up plate   A pattern plate with the cope pattern mounted or attached to one side with the drag on the other.   (*See* Matchplate.)

Bulb sponge   Rubber ball with a small piece of sponge inserted in the hole.

Bumper   Machine for ramming sand in a flask by repeated jarring or jolting action.

Burden   Term used to designate the metal charge for a melting furnace.   It is also used in cost accounting to indicate certain additional charges to be included in assessing costs in the different areas.

Burn-on   Expression denoting adhesion of sand to the casting, usually due to the metal penetrating into the sand.

Burn-out   Usually refers to removal of the disposable wax or plastic pattern in the investment-molding process by heating the mold gradually to a sufficiently high temperature to consume any carbonaceous residues.

Burner   A device which mixes fuel and air intimately to provide perfect combustion when the mixture is burned.   Types include acetylene, oil, gas, powdered coal, stoker, etc.

Bushing   A sleeve, metallic or nonmetallic, usually removable or replaceable, which is placed in a body to resist wear, erosion, etc.

Butt rammer   The flat end of the molder's rammer.

By-product coke   Pulverized coal is placed in sealed ovens 14 to 24 in. wide, 9 to 15 ft high, and 37 to 45 ft long; these ovens are fired externally. The volatile products—gas, ammonia, light oils, and tar generated by the destructive distillation of the coal—are recovered and processed for commercial use.

Calcium-aluminum-silicon   An alloy composed of 10 to 14% calcium, 8 to 12% aluminum, and 50 to 53% silicon, used for degasifying and deoxidizing steel.

Calcium boride   An alloy of calcium and boron corresponding (when pure) to the formula CaB_6, containing about 61% boron and 39% calcium, and used in deoxidation and degasification of nonferrous metals and alloys.

Calcium carbide   A grayish-black, hard crystalline substance made in the electric furnace by fusing lime and coke. Addition of water to calcium carbide forms acetylene and a residue of slaked lime.

Calcium-manganese-silicon   An alloy containing 17 to 19% calcium, 8 to 10% manganese, 55 to 60% silicon, and x0 to 14% iron, used as a cavenger for oxides, gases, and nonmetallic impurities in steel.

Calcium molybdate   A crushed product containing 40 to 50% molybdenum, 23 to 25% lime, 3% iron max., and 5 to 10% silica, used to add molybdenum to iron and steel produced in the open hearth, air furnace, or electric furnace.

Calcium silicon   An alloy of calcium, silicon, and iron, containing 28 to 35% calcium, 60 to 65% silicon, and 6% maximum iron, used as a deoxidizer and degasifier for steel and cast iron. Sometimes called calcium silicide.

Captive foundry   One that is part of a manufacturing plant, and whose products (castings) are used in the plant as parts of finished objects.

Carbide   A compound of carbon with a more positive element, such as iron. Carbon unites with iron to form iron carbide or cementite, Fe_3C.

Carbon boil   Refers to the practice of adding oxidizing agents such as iron ore or oxygen to molten steel in the furnace to react with carbon and create a boiling action. In addition to reducing the carbon content, it removes occluded gases such as hydrogen, oxygen, and nitrogen.

Carbon equivalent   Relationship of total carbon, silicon, and phosphorus in gray iron, expressed by the formula: $CE = TC\% + Si\%/3 + P\%/3$.

Carbon steel   Steel which owes its properties chiefly to various percentages of carbon without substantial amounts of other alloying elements; also known as ordinary steel or straight carbon or plain carbon steel.

Cast iron   Generic term for a series of alloys of iron, carbon, and silicon, in which the carbon is in excess of the amount which can be retained in solid solution in austenite at the eutectic. When cast iron contains a specially added element or elements in amounts sufficient to produce a measurable modification of the physical properties under consideration, it is called alloy cast iron. Silicon, manganese, sulfur, and phosphorus, as normally obtained from raw materials, are not considered as alloy additions.

Cast plate   Metal plate, usually aluminum, cast with the cope pattern on one side and the drag pattern on the other. (*See* Matchplate.)

Cast-weld assembly   Welding one casting to another to form a complete assembly.

Casting (noun)   Metal poured into a mold to form an object.

Casting (verb)   Act of pouring molten metal into a mold.

Casting, machine (verb)   Process of casting by machine.

Casting, open sand (noun)   Casting poured into an uncovered mold.

Casting strains   Strains resulting from internal stresses created during cooling of a casting.

Cavity, mold or die   Impression or impressions in a mold or die that give the casting its shape.

Cement   Mineral substances in finely divided form, which are hardened through chemical reaction or crystallization. A common one is portland cement.

Cement molding   Process in which the sand bonding agent is a type of portland cement that develops high strength early in the hardening stage. Approximately 13 lb of cement, 6 lb of water, and 100 lb of clayfree sand are mixed together. Mixture must be used within 3 to 4 hours. Molds are air dried for 72 hours before use.

Cement, refractory   Highly refractory material in paste or dry form, ready to be mixed with water which may be used as a mortar, a patching material, or to form a complete lining in a furnace or other unit where high temperatures are encountered.

Cementation   Process of introducing elements into the outer layer of metal objects by means of high-temperature diffusion.

Cementite   Iron carbide, Fe_3C, a hard, brittle, crystalline compound observed in the microstructure of iron-base alloys.

Centrifugal casting   Process of filling molds by pouring the metal into a sand or metal mold revolving about either its horizontal or vertical axis, or pouring the metal into a mold that subsequently is revolved before solidification of the metal is complete. Molten metal is moved from the center of the mold to the periphery by centrifugal action.

Ceramic mold   Mold in which the refractory and binder are such that when fired at high temperature, a rigid structure is formed. Mold can be made in a flask or in the form of a shell.

Cereal   Substance derived principally from corn flour, which is added to core and molding sands to improve their properties for casting production.

Cerium   Metallic element, malleable and ductile, most abundant of rare-earth group. Atomic weight 140.13, sp. gr. 7.04, hardness (Moh's) about 2, melting point 640°C. Has exceptionally strong affinity for oxygen, sulfur, hydrogen, nitrogen, etc. Readily decomposes silicates, forming cerium oxide and cerium silicide. Nodularizing agent for some cast iron; also said to "neutralize" effect of some subversive elements when producing nodular cast iron with magnesium additions.

Chalk test   Method of crack detection which consists of applying a penetrating liquid to the part, removing the excess from the surface which is then coated with whiting or chalk. After a short time the penetrant seeps out of the cracks into the whiting, causing an appreciable difference in whiteness.

Chamfer   Breaking or beveling the sharp edge or angle formed by two faces of a piece of wood or other material.

Chamotte   Coarsely graded refractory material prepared from calcined clay and ground firebrick mulled with raw clay, used in steel foundries.

Chaplet   A metallic insert or support to hold a core in position in the mold.

Charcoal (pig) iron   Pig iron reduced in a blast furnace, using charcoal as the fuel.

Charging crane   System for charging the melting furnace with a crane.

Charging door   Opening through which the furnace is charged.

Charging floor   Floor from which the furnace is charged.

Charging machine   Machine for charging the furnace, particularly the open hearth.

Charpy test   A pendulum type of impact test in which a specimen, supported at both ends as a simple beam, is broken by the impact of the falling pendulum. Energy absorbed in breaking the specimen, as determined by the decreased rise of the pendulum, is a measure of the impact strength of the metal.

Cheek   Intermediate sections of a flask inserted between cope and drag. Necessitated by difficulty in molding unusual shapes, or in cases where more than one parting line is required.

Chill (noun)   A metal object placed on the outside or inside a mold cavity to induce more rapid cooling at that point.

Chill (verb)   To cool rapidly.

Chill coating   A material applied to metal chills to prevent oxidation or other deterioration of the surface which might result in blows when molten metal comes in contact with the chills.

Chill coils   Chills made of steel wire formed into helical coils or spirals.

Chill nails   Chills in the form of nails.

Chill test   Method of determining the suitability of a gray iron for specific castings through its chilling tendency, as measured from the tip of a wedge-shaped test bar.

Chill zone   Area of a casting in which chilling occurs, as long sharp edges or exterior corners.

Chilled iron   Cast iron poured against a chill to produce a hard, unmachinable surface.

Chip (verb)   To remove extraneous metal from a casting with hand or pneumatically operated chisels.

Chlorination   A refining or degasification process, wherein dry chlorine gas is passed through molten aluminum-base and magnesium-base alloys to remove entrapped oxides and dissolved gases.

Choke   Restriction in a gating system to control the flow of metal beyond that point.

Chromium   Alloying element used as a carbide stabilizer. (*See* Ferrochromium.)

Chvorinov's rule   A rule which states that solidification time is proportional to the square of the volume of the metal and inversely proportional to the square of the surface area, or t (solidification time) $= KV^2/SA^2$.

Clamp   A device for holding parts of a mold, flask, corebox, etc., together.

Clamp-off   Indentation on a casting surface due to displacement of sand in the mold.

Clay wash   Clay and water mixed to a creamy consistency.

Clay, refractory   A clay which, in addition to its capability of resisting high temperatures, also possesses strong bonding power.

Cleaning   Process of removing sand, surface blemishes, etc., from the exterior and interior surfaces of castings. Includes degating, tumbling or abrasive blasting, grinding off gate stubs, etc.

Coalescence Agglomeration of fine particles into a mass. Also growth of particles of a dispersed phase by solution and reprecipitation. Also grain growth by absorption of adjacent undistorted grains.

Cobalt 60 Radioisotope of the element cobalt used in radiographic examination of castings, and for determining height of molten metal in cupola well.

Coke A porous, gray, infusible product resulting from the dry distillation of bituminous coal which drives off the volatile matter. Used as a fuel in cupola melting. Petroleum coke results from distillation of petroleum, and pitch coke from distillation of coal tar pitch. (*See* Beehive and By-product coke.)

Coke bed First layer of coke placed in the cupola. Also the coke used as the foundation in constructing a large mold in a flask or pit.

Cold blast pig iron Pig iron produced in a blast furnace without the use of the heated air blast.

Cold shortness Brittleness when metal is at a low temperature.

Cold shut Where two streams of metal do not unite thoroughly in a casting.

Collapsibility Tendency of a sand mixture to break down under conditions of casting.

Colloids Finely divided material, less than 0.5 micron (0.00002 in.) in size, gelatinous, highly absorbent, and sticky when moistened.

Columnar structure Coarse structure of parallel columns of grains caused by highly directional solidification resulting from sharp thermal gradients.

Combination die A die-casting die having two or more cavities of dissimilar parts. (*See* Multiple-cavity die.)

Combined carbon The carbon in iron or steel combined with other elements and therefore not in the free state as graphite or temper carbon.

Combustibles Materials capable of combustion; inflammable.

Conductivity The quality or power of conducting or transmitting heat, electricity, etc.

Compressive strength The maximum compressive strength which a material is capable of developing.

Continuous annealing furnace Furnace in which castings are annealed or heat treated by being passed through different zones kept at constant temperatures.

Contraction Act or process of a casting becoming smaller in volume and/or dimensions during the solidification of the metal or alloy which composes the casting.

Controlled cooling Process by which a metal object is cooled from an elevated temperature in a predetermined manner of cooling to avoid hardening, cracking, or internal damage.

Converter Vessel for refining molten metal by blowing air through it. Used in making steel from molten cast iron and in refining copper.

Cooling curve A curve delineating the relationship between temperature and time during the cooling of a metal or alloy test specimen.

Cope The upper or topmost section of a flask, mold, or pattern.

Cope, false A temporary cope which is used only to establish the parting line.

Core Separable part of the mold, usually made of sand and generally baked, to create openings and various shaped cavities in the castings. Also used to designate the interior portion of an iron-base alloy which after case-hardening is substantially softer than the surface layer or case.

Core binder   Any material used to hold the grains of core sand together.

Corebox   Box with an opening in which the core is formed.

Core blowing machine   Machine which rams the core by blowing sand into the corebox.

Core drier (dryer)   Sand or metal supports to keep cores in shape during baking.

Core jig (fixture)   Device in which a number of cores are assembled outside the mold, then used to locate the assembly in the proper position in the mold.

Core machine   Machine for making cores.

Core oil   Linseed-base or other oil used as a core binder.

Core oven   An oven for baking cores.

Core paste   Material in paste form used as an adhesive to join sectional cores.

Core plate   A plate or board made of metal or heat-resisting material on which certain types of cores are baked.

Core print   An extension of the pattern for locating the core or an extension of the mold cavity for locating the core.

Core, ram-up   Core attached to the pattern and rammed up in the mold, where it remains when the pattern is withdrawn.

Core rod   Iron or steel in rod form used to stiffen or support a core internally.

Core sand   Sand for making cores.

Core sand mixer   Equipment in which cores are made.

Core setter   An operator or machine for placing cores in molds.

Core shift   Defect resulting from movement of the core from its proper position in the mold cavity.

Core vents   A wax product, round or oval in form, used to form the vent passage in a core. Also refers to a metal screen or slotted piece used to form the vent passage in the corebox employed in a core blowing machine.

Core wash   Refractory coating for a core.

Core, Washburn   A thin core constricting the riser where it is attached to the casting. It heats quickly, creating a hot spot which prevents temperature drop in metal passing through and promotes feeding of the casting. In many cases it speeds riser removal.

Coremaker   A person who makes cores.

Coring   Variable composition in solid-solution dendrites; the center of the dendrite is richer in one element, as shown by the pertinent solidus-liquidus lines in a phase diagram.

Crack, cold   Appears in a casting after solidification and cooling due to excessive strain generally resulting from nonuniform cooling.

Crack, hot   Developed in a casting before it has cooled completely, and usually due to some part of the mold restraining solid contraction of the metal. (*See* Tear, hot.)

Creep   Time rate of deformation continuing under stress intensities well within the yield point, proportional limit, or the apparent elastic limit for the temperature.

Critical points (temperatures)   Temperatures at which changes in the phase of a metal take place, and are determined by the liberation of heat when the metal is cooled and by the absorption of heat when the metal is heated, resulting in halts or arrests on cooling and heating curves.

Crucible   A ceramic pot or receptacle of graphite-clay, clay, or other refractory material in which metal is melted. This term is sometimes applied to pots of cast iron, cast or wrought steel.

Crucible furnace   Furnace in which metal is melted in crucibles.

Crush   Casting defect appearing as an indentation in the surface due to displacement of sand in the mold; usually at the joint surfaces.

Crystallization   Act or process of forming crystals or bodies formed by elements or compounds solidifying so they are bounded by plane surfaces, symmetrically arranged, and are the external expressions of definite internal structure.

Cupola   Stack-type melting unit in which metal is melted in direct contact with the fuel.

Cupola, basic   Cupola with refractory lining which has a basic reaction, usually magnesite, and is operated with slags high in lime. Lining may be a neutral material like carbon, used with high lime slags.

Cupola blower   A machine which compresses a large volume of air at low pressure for operation of the cupola.

Cupola dust arrester   A device attached to the stack of a cupola which removes dust and sparks from the outgoing gases.

Cupola, hot blast   Cupola in which the air blast is heated to temperatures from 400 to 1000°F.

Cupola, water-cooled   Cupola in which melting zone and tuyeres are cooled with water. Cooling of melting zone may be internally through jackets or steel tubing under the refractory lining. Cooling is also accomplished externally by water flowing down the outer shell.

Cutoff machines, abrasive   A machine using a thin abrasive wheel and employed in cutting off gates and risers from castings or in similar operations.

Cuts   Defects in castings resulting from erosion of the sand by the molten metal pouring over the mold or core surface.

Cutter, gate   A piece of sheet metal or other tool for removing a portion of the sand in a mold to form the gate or metal entrance into the casting cavity.

Cutter, sprue   A piece of metal tubing or other tool used to remove a portion of the sand from a mold to form the sprue or passage from the exterior of the mold to the gate. Also a machine used for shearing sprues and gates from castings.

Damping capacity   The ability to absorb vibration. More accurately defined as the amount of work dissipated into heat by a unit volume of material during a completely reverse cycle of unit stress.

Daub   To coat or plaster the inside of a cupola at the melting zone or the inside of a ladle with a refractory mixture.

Degasifier   A material employed for removing gases from metals and alloys.

Delta iron   An allotropic (polymorphic) form of iron, stable above 2550°F, crystallizing in the body-centered-cubic lattice.

Dendrite   A crystal formed during the solidification of a metal or alloy characterized by a branching structure like that of a fir tree.

Deoxidizer   A material used to remove oxygen or oxides from metals and alloys.

Desulfurizer   A material used to remove sulfur from molten metals and alloys.

Dezincification   Corrosion of some copper-zinc alloys, involving loss of zinc and the formation of a spongy porous copper.

Die-casting (verb)   Pouring molten metal under pressure into metal molds.

Die-casting (noun)   Casting resulting from die-casting process.

Die-casting, cold chamber   Type of casting made in a die-casting machine in which the metal injection mechanism is not submerged in the molten metal.

Die-casting, hot chamber   Type of casting made in a die-casting machine in which the metal injection mechanism is submerged in the molten metal.

Dielectric baking   Baking of cores and molds in a field of high-frequency electric current generated by dielectric equipment; employed with resin-bonded cores.

Diffusion   Movements of atoms within a solution. Net movement is usually from regions of high concentration to regions of low concentration to achieve homogeneity of the solution which may be a solid, gas, or liquid.

Dilatometer   Instrument for measuring expansion or contraction caused by changes in temperature or structure.

Dimensional stability   Ability of a casting to remain unchanged in size and shape under ordinary atmospheric conditions.

Direct-arc furnace   Electric furnace in which the material is heated directly by an arc established between the electrodes and the work.

Directional solidification   Refers to the arrangement of a solidification pattern in a casting by establishment of high temperature gradients, whereby solidification of the metal begins at the point farthest from the metal entrance or sprue and the metal progressively freezes or solidifies to and including the sprue.

Dirt   Indefinite term referring to any extraneous material entering a mold cavity and usually forming a blemish on the casting surface.

Dowel   A short pin of metal, wood, etc., used to join two pieces of material together.

Draft   Taper allowed on the vertical faces of a pattern to permit removal from the sand mold without excessive rapping and tearing of the mold walls.

Drag   The lower or bottom section of a mold or pattern.

Draw bar   A bar used for lifting the pattern from the mold. Also car connection.

Drawback   Section of a mold lifted away on a plate or arbor to facilitate removal of the pattern.

Draw, surface   Appearance of shrink on the upper surface of a casting.

Drop out   Sand falling from the cope of a mold.

Dropping the bottom   Removal of the supporting props under the cupola bottom doors to permit emptying of the remaining contents.

Dross   Metal oxides, etc., on or in a metal or alloy.

Dry permeability   Property of a molded mass of sand dried at 221 to 230°F and cooled to room temperature, to permit passage of gases through it.

Dry-sand mold   A mold made of prepared molding sand dried thoroughly before being filled with metal.

Dry strength   Maximum strength of a sand mixture which has been dried at 105 to 110°C and cooled to room temperature for testing.   Value may be in compression, shear, tensile, or transverse strength.

Ductile iron   Nodular or spheroidal graphite cast iron produced by residual magnesium remaining in the iron after ladle addition of magnesium.

Ductility   The property permitting permanent deformation by stress in tension without rupture.

Duplexing   Term usually used in reference to melting metals or alloys in one type of furnace and transferring to another for holding, refining, etc.   Common in the malleable field, where charges are melted in a cupola and transferred to air or electric furnaces for slight reduction of carbon and an increase in temperature.

Dye penetrant   Penetrant used in crack detection, which has a dye added to make it more readily visible under normal or black-lighting conditions. In the case of normal lighting, the dye is usually red and nonfluorescent. With black lighting, the dye is fluorescent and yellow-green in color.

Ejector marks   Marks left on die-castings by the ejector pins, which may be raised or depressed from the surface of the casting.

Ejector pins   Pins used to eject die-castings from the die.

Ejector plate   Movable plate beneath a shell molding pattern containing the pins for lifting or ejecting the hardened, resin-bonded shell mold from the pattern.

Equilibrium   Dynamic condition of balance between atomic movements where the resultant is zero, a stable condition.

Equilibrium diagram   (*See* Phase diagram.)

Erosion scab   Casting defect occurring where the metal has been agitated, boiled, or has partially eroded away the sand, leaving a solid mass of sand and metal at that particular spot.

Ethyl silicate   Light brown liquid consisting predominantly of tetraethyl silicate with some polysilicates which can be hydrolized with water to form alcohol and silicic acid; the latter in turn dehydrates to an amorphous form of silica extremely resistant to most chemicals and heat.   Used as a bonding agent in investment molding.

Eutectic   The alloy which has the lowest melting point possible for a given composition.

Eutectic reaction   Reaction in which a liquid solution solidifies or transforms at constant temperature to form a solid mass made up of two kinds of crystals.

Eutectoid   A solid solution of any series which cools without change to its temperature of final composition.

Exothermic reaction   A reaction which produces heat.

Expansion scabs   Rough thin layers of metal partially separated from the body of the casting by a thin layer of sand, and held in place by a thin vein of metal.

Expansion, sand   Dimensional increase that sand undergoes when subjected to elevated temperature conditions.

Expendable pattern   In investment molding, the wax or plastic pattern that is left in the mold and later melted and burned out. Also called disposable pattern.

External chills   Various materials of high heat capacity such as metals, graphite, etc., forming parts of the walls of the mold cavity to promote rapid heat extraction from molten metal.

Facing   Refractory material applied to the face of a mold.

Facing sand   Specially prepared sand in the mold adjacent to the pattern to produce a smooth casting surface.

False cheek   A cheek used in making a three-part mold in a two-part mold.

Fatigue crack   A fracture starting from a nucleus where there is an abnormal concentration of cyclic stress, and propagating through the metal. Surface is smooth and frequently shows concentric markings with a nucleus as the center.

Feed head   A reservoir of molten metal provided to compensate for contraction of metal as it solidifies, by the feeding down of liquid metal to prevent voids. Also called a riser.

Ferric oxide   Red iron oxide, Fe_2O_3, commonly available as hematite ore. Used in ground form in cores and molds to increase hot compressive strength.

Ferrite   Iron practically carbon-free. It forms a body-centered-cubic lattice and may hold in solution considerable amounts of silicon, nickel, or phosphorus; hence the term is also applied to solid solutions in which alpha or delta iron is the solvent.

Ferroalloys   Alloys consisting of certain elements combined with iron, and used to increase the amount of such elements in ferrous metals and alloys. In some cases the ferroalloys may serve as deoxidizers.

Ferroaluminum   An alloy of iron and aluminum containing about 20% iron and 80% aluminum.

Ferrochromium   An alloy of iron and chromium available in several grades containing from 66 to 72% chromium and from 0.06 to 7% carbon.

Ferromanganese   An alloy of iron and manganese containing from 78 to 82% manganese.

Ferromolybdenum   An alloy of iron and molybdenum containing 58 to 64% molybdenum.

Ferrophosphorus   An alloy of iron and phosphorus containing about 70% iron and 25% phosphorus.

Ferrosilicon   An alloy of iron and silicon available in several grades containing from 14 to 20% silicon, 42.5 to 52% silicon, 69.5 to 82% silicon, 82 to 88% silicon, and 88 to 95% silicon.

Ferrostatic pressure   Pressure induced by a head of liquid iron or steel.

Fillet   A concave junction formed where two surfaces meet by use of a preformed strip of leather or wax.

Fin   A thin piece of metal projecting from a casting at the parting line or at the junction of cores, or of cores and mold, etc.

Fines   Sand grain sizes substantially smaller than the predominating grain size in a molding sand; also material remaining on 200- and 270-mesh sieves and pan after tests for grain size and distribution.

Finish (machine)   Amount of metal allowed for machining.

Finish (verb)   The hand work on a mold after the pattern has been withdrawn.

Firebrick   Brick made of refractory clay or other material which resists high temperatures.

Fireclay   A type of clay which is resistant to high temperatures.

Flaring   Term used in connection with zinc-bearing alloys, particularly manganese bronze, to denote evolution of zinc oxide fumes during melting.

Flash   Thin fin or web of metal extending from the casting along the joint line as a result of poor contact between cope and drag molds.

Flask   Container in which a mold is made.

Flask pins   Assure proper alignment of cope and drag molds after the pattern is withdrawn.

Flask, slip   A removable flask which can be stripped vertically from the mold.

Flask, snap   A hinged flask which can be removed from the mold after completion.

Flask, tight   Flask which remains on the mold.

Flowability   Property of a foundry sand mixture which enables it to fill pattern recesses and move in any direction against pattern surfaces under pressure.

Fluidity   Ability of molten metal to flow readily; usually measured by the length of a standard spiral casting.

Fluorescent crack detection   Application of penetrating fluorescent liquid to a part, then removing the excess from the surface, which is then exposed to ultraviolet light. Cracks show up as fluorescent lines.

Flux   Any substance used to promote fusion. Also any material which reduces, oxidizes, or decomposes impurities so that they are carried off as slags or gases.

Founding   Art and science of melting and pouring metals and alloys into castings to serve mankind.

Freezing   Term used to denote the solidification process.

Furans   Generic term for a family of chemical compounds including furfural and furfuryl alcohol used as binders for core sands.

Fusion   Change from a solid to a fluid state caused by application of heat.

Gagger (jagger)   An L-shaped rod used for reinforcing sand in the cope mold.

Gamma iron   One of the allotropic (polymorphic) forms of iron which crystallizes in the face-centered-cubic lattice form. When pure, its range of stability is from 2552 to 1670°F.

Gas holes   Rounded cavities caused by generation or accumulation of gas or entrapped air in a casting; holes may be spherical, flattened, or elongated.

Gate   Specifically, the point at which molten metal enters the casting cavity.   Sometimes employed as a general term to indicate the entire assembly of connected columns and channels carrying the metal from the top of the mold to that part forming the casting cavity proper. Term also applied to pattern parts which form the passages, or to the metal that fills them.

Gated patterns   One or more patterns with gating systems attached.

Gilsonite   Natural black lustrous asphalt found in the Uinta mountains in Utah and also known as uintaite.   It is used as a carbonaceous addition to foundry sands.

Gooseneck   The pressure vessel or metal injection mechanism in a hot-chamber-type die-casting machine.

Grain fineness number (AFS)   (*See* AFS fineness number.)

Grain refiner   Any material added to a liquid metal or alloy or a treatment which produces a finer grain size in the subsequent solid.

Grains   Crystals in metals and alloys.

Granular pearlite   A structure formed from ordinary lamellar pearlite by long annealing at a temperature below but near to the critical point, causing the cementite to spheroidize in a ferrite matrix.

Graphite   Native carbon in hexagonal crystals, also foliated or granular massive, of black color with metallic luster, and soft. Used for crucibles, foundry facings, lubricants, etc. Also made artificially by passing alternating current through a mixture of petroleum coke and coal tar pitch.

Graphite, primary   Carbon precipitated as graphite flakes while the iron cools through the freezing eutectic in which austenite, graphite, molten iron, and carbide exist together.   Usually with reference to white fracture cast iron.

Graphite, secondary   Graphite formed by decomposition of austenite during slow cooling of cast iron.

Graphitization   The decomposition of carbide to give free carbon as graphite or as temper carbon.

Graphitizer   Any substance, such as silicon, titanium, aluminum, etc., which promotes the formation of graphite in cast iron compositions.

Green permeability   Property of a molded mass of sand in its tempered condition which is a measure of its ability to permit the passage of gases through it.

Green sand   Prepared molding sand in the moist or as-mixed condition.

Green sand core   Core used in the green state; not baked.

Green strength   Tenacity (compressive, shear, tensile, or transverse) of a tempered sand mixture.

Grinding   Removing gate stubs, fins, and other projections on castings by an abrasive wheel.

Growth   With reference to cast iron, permanent increase in volume that results from continued or repeated cyclic heating and cooling at elevated temperatures.   For unalloyed iron, temperature is in excess of 900°F, and growth is caused by decomposition or graphitization of carbides and by oxidation of the graphite.

Guide pin   The pin used to locate the cope in the proper place on the drag.

Gypsum cement   Calcined calcium sulfate, commonly called plaster of Paris.

Hand ladle or shank   A small ladle carried by one man.

Hard sand match (matchplate)   A body of sand shaped to conform to the parting line upon which a pattern is laid in starting to make a mold. Sand is made hard by addition of linseed oil and litharge, portland cement, etc. (*See* Match.)

Head   Pressure exerted by a fluid such as molten metal. Also used as a term for a riser.

Heap sand   Sand in piles on the foundry floor.

Hearth   That portion of a reverberatory furnace on which the molten metal or bath rests.

Heat   A stated tonnage of metal obtained from a period of continuous melting in a cupola or other furnace.

Heat transfer   Transmission of heat from one body to another by radiation, convection, or conduction.

Holding furnace   Usually a small furnace for maintaining molten metal at the proper pouring temperature, and which is supplied from a large melting unit.

Holding ladle   Heavily lined and insulated ladle in which molten metal is placed until it can be used. (*See* Holding furnace.)

Horn gate   Curved gate in the shape of a horn arranged to permit entry of molten metal at the bottom of the casting cavity.

Hot deformation (sand)   Change of form of a sand specimen which accompanies the determination of hot strength.

Hot shortness   Brittleness in metal at elevated temperature.

Hot spots   Term applied to gray iron castings to denote chilled areas or inclusions that are harder than the surrounding iron and that cause machining difficulties.

Hot strength (sand)   Tenacity (compressive, shear, tensile, or transverse) of a sand mixture determined at an elevated temperature.

Hot tears   Cracks in castings formed at elevated temperatures; usually by contraction stresses.

Hypereutectic alloy   An alloy containing more than the eutectic amounts of the solutes. Analogous to hypereutectoid.

Hypereutectoid   An alloy containing more than the eutectoid composition.

Hypoeutectoid   An alloy containing less than the eutectoid composition.

Impregnation   A process for salvaging leaky castings by injecting under pressure liquid synthetic resins, tung oil, etc., into the porous area. This material is then solidified in place by heating or baking.

Impression   Cavity in a die-casting die or in a mold.

Inclusions   Particles of slag, sand, or other impurities such as oxides, sulfides, silicates, etc., trapped mechanically during solidification or formed by subsequent reaction of the solid metal.

Indirect-arc furnace   Electric furnace in which the arc is struck between two horizontal electrodes, heating the metal charge by radiation.

Induction furnace   A melting unit wherein the metal charge is melted electrically by induction.

Ingot   Commercial pig mold or block in which copper, copper-base, aluminium, aluminium alloys, magnesium, magnesium alloys, and other nonferrous materials are made available to the foundry man.

Injection   Forcing molten metal into a die-casting die. Also refers to forcing oxygen, nitrogen, and other gases, as well as solids such as calcium carbide and graphite, into molten metal.

Inoculation   A process of adding some material to molten metal in the ladle for the purpose of controlling the structure to an extent not possible by control of chemical analysis and other normal variables.

Insulating sleeve   Hollow cylinders or sleeves formed of gypsum, diatomaceous earth, pearlite, vermiculite, etc. Placed in the mold at sprue and riser locations to decrease heat loss and rate of solidification of the metal contained in them.

Internal chills   Solid pieces of metal or alloy, similar in composition to the casting, placed in the mold prior to filling it with molten metal. They increase the rate of solidification in their areas and are employed only where feeding is difficult or impossible.

Internal shrinkage   Void or interconnected voids appearing in the interior of a casting; caused by improper or insufficient feeding during the solidification process.

Internal stress   A system of balanced forces existing within a casting not subjected to a working load.

Inverse chill   A condition in an iron casting section in which the interior is mottled or white while the outer sections are gray. Also called reverse chill, internal chill, or inverted chill.

Investment molding   Method of molding using a pattern of wax, plastic, or other material which is "invested" or surrounded by a molding medium in slurry or liquid form. After the molding medium has solidified, the pattern is removed by subjecting the mold to heat, leaving a cavity for reception of molten metal. Also called lost-wax process or precision molding.

Iron, malleable   A mixture of iron and carbon, including smaller amounts of silicon, manganese, phosphorus, and sulfur, converted structurally by heat treatment into a matrix of ferrite containing nodules of temper carbon.

Iron, white or hard   Iron of suitable composition in which the castings, later to be malleableized, are originally cast. Carbon is in the combined form; hence its white fracture and name.

Jacket, mold   A wood or metal form slipped over a mold made in a snap or slip flask, to support the four sides of the mold during pouring. Jackets and mold weights are shifted from one row of molds to another during the pouring period.

Jobbing foundry   Foundry which is not a part of a manufacturing plant, and produces castings for sale. Usually makes a wide variety of castings in small lots or quantities.

Jolt machine   Molding machine which packs or rams the sand around the pattern by raising the table on which the flask, sand, and pattern are mounted a few inches and allowing the whole to drop suddenly. The table is raised pneumatically, and the operation is repeated until the desired sand density is reached.

Jolt-squeeze machine Combination molding machine on which the sand is rammed into the flask by jolting (*see* Jolt machine), then compressed further by a mechanism that uses fluid pressure to force the table and contained flask upward against a fixed plate. The plate is slightly smaller in dimension so that it fits inside the flask.

Kaolinite Hydrated silicate of alumina represented by the formula $Al_2O_3 \cdot 2\ SiO_2\ 2H_2O$. It is a white, pearly mineral, crystallizing in a monoclinic system in the form of small, hexagonal plates. Constituent of kaolin, white china clay, used for porcelain, etc.

Killed steel Molten steel held in a ladle, furnace, or crucible (and usually treated with aluminum, silicon, or manganese) until more gas is evolved and the metal is perfectly quiet.

Kish Graphite thrown out by liquid cast iron in cooling.

Knock out To remove sand and casting from a flask.

Knock-out pins Small pins on die-casting machines, permanent molds, and shell-molding machines for ejection of castings, etc. (*See* Ejector pins.)

Ladle Metal receptacle lined with refractory for transportation of molten metal. Types include hand, bull, sulky, trolley, crane, bottom-pour, and teapot.

Ladle, bull Large ladle for transporting and pouring molten metal.

Lance, oxygen Long steel pipe or tube, usually covered with refractory, used to inject oxygen into molten steel to reduce the carbon content. Also may be used to open up frozen tapholes in cupolas, etc.

Leaker Foundry term for castings which leak under liquid or gaseous pressure.

Lining Inside refractory layer of firebrick, clay, sand, or other material in a furnace or ladle.

Linseed oil Drying-type oil expressed from flax seeds and used as a binder for core sand.

Liquid contraction Shrinkage or contraction in molten metal as it cools from one temperature to another while in the liquid state.

Liquidus The temperature at which solidification of metal begins on cooling and the temperature at which the last portion of solid metal becomes liquid on heating.

Loam A course, strongly bonded molding sand used for loam and dry-sand molding.

Loam molding A system of molding, especially for large castings, wherein the supporting structure is constructed of brick. Coatings of loam are applied to form the mold face.

Loose piece Part of a pattern so attached that it remains in the mold, and is removed after the body of the pattern is drawn. In die-casting, a type of core (which forms undercuts) positioned in, but not fastened to, a die and so arranged as to be ejected with the die-casting, from which it is removed and used repeatedly for the same purpose.

Machinability The capability of being cut, turned, broached, etc., by machine tools.

Magnaflux Trade name for a method of magnetic crack detection.

Magnaglo Trade name for a method of magnetic crack detection in which the magnetic particles are treated so that they fluoresce in ultraviolet light.

Magnetic crack detection   Method of locating cracks in materials which can be magnetized; done by applying magnetizing force and applying finely divided iron powder which then collects in the region of the crack.

Malleability   The property of being permanently deformed by compression without rupture.

Malleableization   Annealing or heat-treating operation performed on white iron castings to transform the combined carbon into temper carbon.

Manganese   One of the elements; its chemical symbol is Mn. Its formula weight is 54.93; specific gravity 7.2, and melting point 1260°C. Metallic manganese is used in the nonferrous industry both as a deoxidizing agent and as an essential constituent to improve physical properties of certain alloys.

Manganese briquets   Crushed ferromanganese bonded with a special refractory in briquet form, and containing 2-lb metallic manganese and $\frac{1}{2}$-lb metallic silicon.

Master pattern   The pattern from which the working pattern is cast.

Match   A form of wood, plaster of Paris, sand, or other material on which an irregular pattern is laid or supported while the drag is being rammed.

Matchplate   A metal or other plate on which patterns split along the parting line are mounted back to back with the gating system to form an integral piece.

Melting pot   Metal, graphite-clay, or ceramic vessel in which metal is melted.

Melting range   Pure metals melt at one definite temperature, but constituents of alloys melt at different temperatures, and the variation from the lowest to the highest is called the melting range. Example is copper, which melts at 1981°F, and the 85-5-5-5 alloy, which melts at 1568 to 1849°F.

Melting rate   Amount of metal melted in a given period of time, usually one hour.

Melting zone   Portion of the cupola above the tuyeres in which the metal melts.

Metal penetration   Defect in the casting surface which appears as if the metal has filled the voids between the sand grains without displacing them.

Metallurgy   Science dealing with the constitution, structure, and properties of metals and alloys, and the processes by which they are obtained from ore and adapted to the use of man.

Microporosity   Extremely fine porosity in castings caused by shrinkage or gas evolution and apparent on radiographic film as mottling.

Microradiography   Process of passing x-rays through a thin section of an alloy in contact with photographic film, and then magnifying the radiograph 50 to 100 diameters to observe the distribution of alloying constituents, of voids, and of other microstructural features.

Microstructure   The structure and characteristic condition of metals as revealed on a ground and polished (etched or unetched) specimen at magnifications above 10 diameters.

Mischmetal   Alloy of rare-earth metals containing about 50% cerium and 50% lanthanum, neodymium, and similar elements.

Misrun   A casting not fully formed.

Mold   The form, usually of sand, containing the cavity into which molten metal is poured to make the casting.

Mold cavity   Impression left in the sand mold by the pattern.

Mold clamps   Devices used to lock or hold cope and drag together.

Mold conveyor   Power-driven unit on which molds are conveyed from the molding station to pouring station to shakeout.

Mold hardener   In sand molds in which sodium silicate is the binder, injection of CO_2 causes a chemical reaction which results in a rigid structure.

Mold oven   Oven or furnace in which molds are dried.

Mold shift   Casting defect resulting when the two cavities in the cope and drag molds do not match properly.

Mold wash   Usually an aqueous emulsion, containing various organic or inorganic compounds or both, which is used to coat the face of a mold cavity. Materials include graphite, silica flour, etc.

Mold weights   Weights placed on top of molds to offset internal or ferrostatic pressure.

Molding machine   Hand or pneumatically operated machine on which molds are made and which rams the sand by squeezing or jolting or both.

Molding sand   Mixture of sand and clay suitable for mold making.

Molding, bench   Making sand molds from loose or production patterns at a bench.

Molding, floor   Making sand molds from loose or production patterns at a floor. Patterns usually are too large to be handled satisfactorily on the bench.

Molding, machine   Making sand molds from production patterns on molding machines.

Montmorillonite   A mineral with the formula $(MgCa)O \cdot Al_2O_3 \cdot 5\ SiO_2 \cdot NH_2O$; it is the chief constituent of bentonite, which is used as a sand binder.

Mottled   White iron structure interspersed with spots or flecks of gray.

Muller   Type of foundry-sand-mixing machine.

Multiple-cavity die   A die-casting die having more than one impression of the same part. (*See* Combination die.)

Multiple mold   Composite mold made up of stacked sections. Each section produces a complete gate of castings. All castings are poured from a central downgate.

Nail, chill   Steel nail with a heavy head which is inserted in the mold wall to hasten cooling of the metal at that point.

Natural sand   Generic term used to describe claybonded sands, suitable for molding operations to produce castings; widely distributed except for the Western section of the U.S.

Nickel   One of the elements; its chemical symbol is Ni. Its formula weight is 58.69; specific gravity 8.90, and melting point 1452°C.

Nodular graphite   Graphite or carbon in the form of spheroids.

Nodular iron   Cast iron which has the major part of its graphitic carbon in nodular form. (*See* Ductile iron.)

Nucleation   Initiation of a phase transformation at discrete sites; the new phase grows on nuclei. (*See* Nucleus.)

Nucleus   The first structurally determinate particle of a new phase or structure that may be about to form. Applicable in particular to solidification, recrystallization, and transformations in the solid state.

Oil core   A core bonded with oil.

Oil furnace   Furnaces fired with oil.

Olivine   Magnesium-iron-orthosilicate composed of forsterite and fayalite. Does not contain free silica. Possible molding material.

One-piece pattern   Solid pattern.

Open-hearth furnace   A refractory-lined, shallow-bath, rectangular furnace in which both hearth and charge are subjected to the direct action of the fuel flame. Fuel may be producer gas, coke-oven gas, powdered coal, or oil. Flame is created by mixing preheated air with fuel in ports. Air is preheated in regenerators called checker chambers.

Open-hearth steel   Steel made in open-hearth furnace.

Open riser   A riser open to the atmosphere. Compare with blind riser.

Open sand casting   A casting poured in a mold which has no cope or other covering.

Optical pyrometer   A temperature-measuring device through which the observer sights the heated object and compares its incandescence with that of an electrically heated filament whose brightness can be regulated.

Overflow well   A recess in a die-casting die connected to the die cavity and functioning as a vent.

Oxidation   Any reaction whereby an element reacts with oxygen.

Oxidizing flame   A flame produced with excess oxygen.

Pad (padding)   Metal added deliberately to the cross section of a casting wall, usually extending from a riser, to ensure adequate feeding to a localized area in which a shrink might occur without the addition.

Parting   A dividing line at which sections of a mold are separated.

Parting compound   Material dusted or sprayed on a pattern or mold to prevent adherence of sand.

Parting line   The line along which a pattern is divided for molding, or along which the sections of a mold or die separate.

Pattern   Model of wood, metal, plaster, or other material used in making a mold.

Pattern coating   Coating material applied to wood patterns to protect them against moisture and abrasion of molding sand.

Pattern letters   Metal or plastic letters and figures in various sizes which are affixed to patterns for identification purposes.

Pattern plates   Straight flat metal or other plates on which patterns are mounted.

Pattern, split   Pattern usually made in two parts, sometimes in more than two.

Patternmaker's shrinkage   Shrinkage allowance made on all patterns to compensate for the change in dimensions as the solidified casting cools in the mold from freezing temperature of the metal to room temperature. Pattern is made larger by the amount of shrinkage characteristic of the particular metal in the casting and the amount of hindered contraction to be encountered. Rules or scales are available for use.

Pearlite A microconstituent of iron and steel consisting of alternative layers of ferrite and iron carbide or cementite.

Pearlitic malleable iron Irons made from the same or similar chemical compositions as regular malleable iron, but so alloyed or heat treated that some of the carbon in the resultant material is in the combined form.

Peen Small end of a molder's rammer.

Pencil core Small cylindrical core used with Williams or atmospheric riser (see which).

Permanent mold A long-life mold into which metal is poured by gravity.

Permeability The property in sand molds which permits the passage of gases. Magnetic permeability of a substance is the ratio of the magnetic induction of the substance to the magnetizing field to which it is subjected.

pH The negative logarithm of the hydrogen ion activity. It denotes the degree of acidity or basicity of a solution. At 25°C, the neutral value is 7. Acidity increases with decreasing values below 7, and basicity increases with increasing values above 7.

Phase A constituent which is completely homogeneous, and is both physically and chemically separated from the rest of the alloy by definite bounding surfaces; for example, austenite, ferrite, cementite. Not all constituents are phases; pearlite, for example.

Phase diagram Graphical representation of the equilibrium temperatures and the composition limits of phase fields and phase reactions in an alloy system.

Phosphorus One of the elements; its chemical symbol is P. Its formula weight is 123.92; specific gravity 1.82, and melting point 44.1°C.

Pig iron Product of the blast furnace by the reduction of iron ore. Also the overiron in the foundry poured into pig molds.

Pilot casting Usually the first casting made from a production pattern and examined for dimensional accuracy, quality, and other features before the pattern is placed on the line.

Pinhole Small hole under the surface of a casting.

Pins, flask Hardened steel locating pins used on flasks to ensure proper register of cope and drag molds.

Pipe Cavity formed by contraction in metal during solidification of the last portion of liquid metal, as in a riser.

Pit mold Mold in which the lower portions are made in a suitable pit or excavation in a foundry floor.

Pitch Usually coal-tar pitch obtained in manufacture of coke and distilled off at about 350°F. Used as a binder in large cores and molds. Melting range is 285 to 315°F.

Plaster moulding Molding method wherein gypsum or plaster of Paris is mixed with fibrous talc, with or without sand, and with water to form a slurry that is poured around a pattern. In a short period of time, the mass air-sets or hardens sufficiently to permit removal of the pattern. The mold so formed is baked at elevated temperature to remove all moisture prior to use. One variation is the Antioch process.

Plastic pattern   Pattern made from any of the several thermosetting-type synthetic resins such as phenol formaldehyde, epoxy, etc.   Small patterns may be cast solid, but large ones are usually produced by laminating with glass cloth.

Plates, bottom   Plates, usually of metal, on which molds are set for pouring.

Plates, core drying   Straight, flat plates of metal or heat-resisting composition on which cores are placed for baking.

Pneumatic tools   Grinders, rammers, drills, etc., operated by compressed air.

Porosity   Unsoundness in castings appearing as blowholes and shrinkage cavities.

Pot   Term usually applied to cast iron containers used in melting aluminum-base alloys; also used to describe steel crucibles for melting magnesium-base alloys, as well as graphite crucibles.

Pour   Discharge of molten metal from the ladle into the mold.

Poured short   Casting which lacks completeness due to the cavity not being filled with molten metal.

Pouring basin   Reservoir on top of the mold to receive molten metal.

Pouring cup   Article made of sand or ceramics containing a cup-shaped depression which is placed over a sprue opening and acts as a funnel to receive the metal poured from the ladle. (*See* Pouring basin.)

Pouring device   Mechanically operated device with a ladle set for controlling the pouring operation.

Pouring ladle   Ladle used to pour metal into the mold.

Powdered coal   Finely ground, high-volatile coal used for heating furnaces and annealing ovens in the malleable foundry industry.

Preheating   General term for a heating which is applied preliminary to some further thermal or mechanical treatment.

Print back   After the surface of a mold is dusted with graphite facing, the pattern is replaced, rapped into position and again removed.

Production foundry   Highly mechanized foundry for manufacturing large quantities of repetitive castings.

Progressive solidification   (*See* Directional solidification.)

Purifiers, flux   Various materials added to molten metals and alloys for the purpose of removing impurities, gases, etc.

Push-up   An indentation in the casting surface due to displacement (expansion) of the sand in the mold.

Pyrometer   An instrument for determining elevated temperatures.

Radiographic inspection   Examination of the soundness of a casting by study of radiographs taken in various areas or of the whole casting.

Radiographic testing   Use of x- or gamma rays in studying the internal structure of objects to determine their homogeneity.

Ram   To pack the sand in a mold.

Ram-up core   (*See* Core, ram-up.)

Rammer   Tool for ramming the sand.

Rapping bar   A pointed bar (or rod) made of steel or other metal, which is inserted vertically into a hole in a pattern, or driven into it, then struck with a hammer on alternate sides to cause vibration and loosening of the pattern from the sand.

Rapping plate   Metal plate attached to a pattern to permit rapping for removal from the sand.

Rat tail   Minor sand buckle occurring as a small irregular line or series of lines.

Rebonding   Term usually employed in reference to adding new bonding material to used molding sand so that it can be used again to produce molds.

Reducing flame   Flame burning with insufficient oxygen to provide complete combustion, resulting in the presence of carbon in the flame.

Refractory   Material usually made of ceramics, which is resistant to high temperatures, molten metal, and slag attack.

Relief sprue   The term usually refers to a second sprue at opposite end of the runner to relieve pressure created during pouring operation.

Resin binder   Any of the thermosetting types of resins used as binders for producing cores and shell molds, such as phenol and urea formaldehydes, melamines, furans (fufuryls and furfuryl alcohol), etc.

Resin-coated sand   Molding or core sand in which the binder is resin applied to the sand as a coating by either cold or hot coating.

Reverberatory furnace   Melting unit with a roof arranged to deflect the flame and heat toward the hearth on which the metal to be melted rests.

Reynolds number   The ratio of the inertia forces in a flowing fluid to the viscous forces. Inertia force is the product of mass and acceleration, and viscous force is equal to the shear stress multiplied by the area.   Hence Reynolds number

$$R_n = \frac{\rho v d}{\mu},$$

where $\rho =$ density of the fluid, $v =$ mean velocity of flow, $d =$ diameter of the channel, and $\mu =$ viscosity of the fluid.

Riddle   Hand- or power-operated device for removing large particles of sand or foreign material from foundry sand.

Rigging   Equipment used for making a mold.

Riser   Reservoir of molten metal attached to the casting to compensate for the internal contraction of the casting during solidification.

Riser gating   Gating system in which molten metal from the sprue enters a riser close to the mold cavity and then flows into the mold cavity.

Rockwell hardness testing   Method of determining the indentation hardness by measuring the depth of residual penetration by a steel ball or a diamond cone.

Rolling over   Operation of reversing the position of the mold so that the pattern faces upward in order to be removed.

Rollover board   Wood or metal plate on which the pattern is laid face down for ramming of the mold.

Rollover machine   Molding machine on which the mold is rolled over before the pattern is drawn.

Run-out   Metal flowing through a defect in the mold.

Runner   The portion of the gate assembly which connects the downgate or sprue with the casting.

SG iron   Term used in Britain and continental Europe for ductile or nodular iron. SG means spherulitic or spheroidal graphite.

Sag   Defect which appears as an increase or decrease in metal section due to sinking of sand in the cope (decreased section) or sagging in the core (increased section).

Sand blast   Sand driven by a blast of compressed air (or steam). It is used to clean castings, to cut, polish, or decorate glass or other hard substances, and also to clean building fronts, etc.

Sand castings   Metal castings produced in sand molds.

Sand conditioning   Preparation of used molding sand for reuse, which includes additions of bond, additives, moisture, etc.

Sand control   Procedure used to adjust various properties of sand such as fineness, permeability, green strength, moisture content, etc., in order to obtain castings free from such defects as blows, scabs, rat tails, veins, etc.

Sand control equipment   Testing instruments such as moisture determinators, permeability air-flow apparatus, etc., for determining the various physical properties of sands.

Sand dryer   Apparatus for removing moisture from sand.

Sand holes   Cavities of irregular shape and size whose inner surfaces plainly show the imprint of granular material.

Sand muller   A machine for mixing sand by kneading and squeezing.

Sand reclaimer   Equipment for removing extraneous material from used sand and reconditioning it for further use.

Sand slinger   Molding machine which throws sand into a flask or corebox, by centrifugal action.

Sand tempering   Adding sufficient moisture to core or molding sand to make it workable.

Sand toughness   Indication of molding sand workability, particularly with reference to rammability, because the tougher the sand, the harder it is to ram tightly against the pattern. It is usually given as a number obtained by multiplying deformation by green compressive strength times 1000.

Sand, backing   Sand in a mold back of the facing.

Sand, bank   Sand from a bank or pit.

Sand, blast   Sand used in an abrasive blasting machine for cleaning castings.

Sand, core   Sand used in making cores.

Sand, facing   Prepared sand used next to the pattern.

Sand, floor   Sand used in floor molding.

Sand, heap   Sand prepared on foundry floor.

Sand, lake   Sharp sand from vicinity of lakes.

Sand, molding   Sand used to make molds.

Sand, natural   Naturally bonded sand as distinguished from that which is formed synthetically.

Sand, open   Sand through which gases can pass freely.

Sand, silica   Sand composed of almost pure silica.

Sand, synthetic   Molding sand prepared by adding clay or other bond to the sand which is practically free of those materials.

Scab   A blemish on a casting caused by eruption of gas from the mold face.

Sea coal   Finely ground bituminous coal.

Seam   Surface defect on a casting similar to a cold shut, but not as severe.

Segregation   Concentration of alloying elements at specific regions, usually as a result of the primary crystallization of one phase with the subsequent concentration of other elements in the remaining liquid.

Shakeout   The operation of removing castings from the mold.

Shakeout machinery   Equipment for mechanical removal of castings from molds.

Shank   The handle attached to a small ladle.

Sharp sand   Sand that is substantially free of bond; the term does not refer to grain shape.

Shear strength   Maximum shear stress which a material can develop.

Shell molding (croning process)   Process in which clay-free silica sand coated with the thermosetting resin or mixed with the resin is placed on a heated metal pattern for a short period of time to form a partially hardened shell. The unaffected sand mixture is removed for further use. The pattern and the shell are then heated further to harden or polymerize the resin-sand mix, and the shell is removed from the pattern.

Shift   A casting defect resulting from a mismatch of cope and drag.

Shot   Abrasive blast cleaning material. In die-casting, it is the phase of the die-casting cycle when molten metal is forced into the die.

Shrink hole   A hole or cavity in a casting resulting from contraction and insufficient feed metal, and formed during the time the metal changed from the liquid to the solid state.

Shrink rule   Patternmaker's rule graded to allow for metal contraction.

Shrinkage, centerline   Shrinkage occurring in the center of casting sections, particularly with platelike or barlike contours, which solidify simultaneously from two faces and cut off feeding in the central portion.

Sieve   A device with meshes of wire or other material for separating fine material from coarse material.

Silica   Silicon dioxide, SiO_2, occurring in nature as quartz, opal, etc. Molding and core sands are impure silica.

Silica flour   Silica in finely divided form.

Silica wash   Silica flour mixed with water and other materials to form a brushable or sprayable facing material.

Silicon   One of the elements with the chemical symbol Si whose formula weight is 28.06, specific gravity 2.4, and melting point 1420°C.

Silicon-aluminum   An alloy of 50% silicon and 50% aluminum used for making silicon additions to aluminum alloys; also called an intermediate or hardener alloy. Melting point is 1070°F.

Silicon brass   A series of alloys containing 0.5–6% silicon, 1–19% zinc, and a substantial amount of copper.

Silicon bronze   A series of alloys containing 1–5% silicon, 0.5–3% iron, under 5% zinc, under 1.5% manganese, and the remainder being substantially copper.

Silicon carbide briquets   Silicon carbide in briquet form used as an inoculant and deoxidizer in cupola-melted gray iron.

Silicon-copper   An alloy of silicon and copper, used as a deoxidizer and hardener in copper-base alloys, which is available in two types containing 10 and 20% silicon. The 10% grade has a melting point of 1500°F while the 20% grade melts at 1650°F.

Silvery iron   A type of pig iron containing 8–14% silicon, 1.50% carbon max., 0.06% sulfur max., and 0.15% phosphorus max.

Sintering point   The temperature at which a molding material begins to adhere to a casting, or, in a test, the point when the sand coheres to a heated platinum ribbon under controlled conditions.

Skeleton pattern   A pattern made in outline to reduce cost.

Skim bob   Small upward bulge in the gating system, near the casting cavity, which functions as a dirt trap.

Skim core   Flat core or tile placed in a runner system to skim the flowing stream of metal. In a pouring basin it holds back the slag and dross, permitting clean metal to pass underneath.

Skim gate   An arrangement which changes the direction of flow of molten metal in the gating system and thereby prevents the passage of slag and other extraneous materials beyond that point.

Skimmer   A device or tool for removing slag and dross from surface of molten metal.

Skin   The surface of a mold or casting.

Skin drying   Drying of the mold surface by direct application of heat.

Slag   A nonmetallic covering on molten metal as the result of the combining of impurities contained in the original charge, such as ash from the fuel, and any silica and clay eroded from the refractory lining. Except in bottom pour ladles, it is skimmed off prior to pouring the metal.

Slurry   Thin watery mixture such as the gypsum mixture for plaster molding, the molding medium used in investment molding, core dips, and mold washes.

Slush casting   Casting made by pouring an alloy into a metal mold, allowing it to remain sufficiently long to form a thin solid shell, and then pouring out the remaining liquid metal.

Smelter   An individual or firm which wins metals from ores, or which melts, treats or refines scrap metals and alloys for further use.

Snag   Removal of fins and rough places on a casting by means of grinding.

Soldiers   Thin pieces of wood used to strengthen a body of sand or hold it in place.

Solid contraction   Shrinkage or contraction as a metal cools from the solidifying temperature to room temperature.

Solid solution   A single solid homogeneous crystalline phase containing two or more chemical species.

Solidification   Process of a metal (or alloy) changing from the liquid to the solid state.

Solidification range   Only pure metals solidify or freeze at one definite temperature. Alloys contain different constituents which solidify at different temperatures, and the various temperatures from that of the first constituent to solidify to that of the last constituent to freeze is called the solidification range.

Solidifying contraction   Shrinkage or contraction as a metal solidifies.

Solidus   Temperature at which freezing is completed. Below that temperature all metals are completely solid.

Spectograph   Optical instrument for determining the concentration of metallic constituents in a metal (or alloy) by the intensity of specific wave lengths generated when the metal or alloy is thermally or electrically excited.

Spiegeleisen (spiegel)   A high manganese pig iron containing 15–30% manganese and used in Bessemer and open-hearth steel production.

Sprue   The vertical portion of the gating system where the molten metal first enters the mold. In die-casting, the metal that fills the conical passage (sprue hole) connecting the nozzle with runners.

Sprue base   The lower end of the sprue attached to the runner system. It is usually in the form of an enlargement or reservoir to reduce turbulence.

Sprue button   A device attached to the cope pattern to indicate where the sprue should be cut.

Sprue cutter   A piece of tubing which cuts the sprue hole through the cope. Also a shear-type machine for removing the sprue and gates from the casting.

Sprue pin   In die-casting, a tapered pin with a rounded end projecting into a sprue hole, acting as a core that deflects the metal and aids in removal of the sprue from the die-casting.

Sprue plug   A tapered metal or wood pin used to form the sprue opening in a mold. Also a metal or other stopper used in a pouring basin to prevent molten metal from flowing into the sprue until a certain level has been reached. It prevents entry of dirt and dross.

Spruing   Removing gates and risers from castings after the metal has solidified.

Squeeze pressure   The pressure applied by a molding machine to press the flask and contained sand against the fixed squeeze head or board on a molding machine.

Stack molding   Molding method in which the half-mold forms the cope and drag. They are placed one on top of the other and poured through a common sprue. Cavities on the bottom side of one half-mold rest on the flat side of the half-mold beneath. When the cavities are in both sides of the half-molds, the method is called multiple molding.

Step gate   A vertical sprue containing a number of side branches or entries at different levels into the casting cavity.

Stop off   To shorten or change a mold.

Stop off strip   Reinforcing members on frail patterns. Impressions later filled with sand.

Strained castings   Molten metal, when poured into the mold too fast, raises the cope slightly from the drag and produces an oversize casting with protruding fins; an oversize casting can also be produced from a weak mold.

Strainer core   A perforated core placed at the bottom of a sprue or in other locations in the gating system to control the flow of the molten metal. To some extent, it prevents coarse particles of slag and dross from entering the mold cavity.

Strains, casting   Strains produced by internal stresses, resulting from unequal contraction of the metal as the casting cools.

Strike off   A straight edge to cut the sand level with the top of the drag or cope flask.

Stripping machine   A device for removing the pattern from a mold or a core from the corebox.

Sulfur   A chemical element having symbol S, formula weight 32.06, specific gravity 2.046, and melting point 120°C.

Supercooling (undercooling)   Cooling below the temperature at which an equilibrium phase transformation can take place without actually obtaining the transformation.

Superheating   Theoretically, the temperature above the liquidus; in practice, it usually means temperature above the usual pouring range.

Surface finish   Condition or appearance of the surface of a casting.

Sweep   To form a mold or core by scraping the sand with a form sweep having the desired profile.

Synthetic sand   Any sand compounded from selected individual materials which, when mixed together, produce a mixture of proper physical properties from which to make molds.

Tap   To withdraw a molten charge from the melting unit.

Taphole   Opening in a furnace through which molten metal is tapped into the forehearth or ladle.

Teapot ladle   Ladle with an external spout wherein the molten metal is poured from the bottom rather than from the top.

Tear, hot   Same meaning as hot crack, but developing before the casting has solidified completely.

Temper carbon   Carbon in nodular form, characteristic of malleable iron.

Tempering (sand)   Addition of water to and mixing molding sand to obtain uniform distribution of moisture.

Tensile strength   The greatest load per square inch of original cross-sectional area carried during a tension test.

Ternary alloy   One which contains three principal elements.

Thermal contraction   Decrease in linear dimensions of a material which accompanies a change in temperature.

Thermal expansion   Increase in linear dimensions of a material which accompanies a change in temperature.

Thermit reaction   Exothermic, self-propagating processes in which finely divided aluminum powder is used to reduce metal oxides to free metals by direct oxidation of aluminum to aluminum oxide, with accompanying reduction of the less stable metal oxide.

Thermocouple   A bimetallic device capable of producing an electromotive force roughly proportional to temperature differences on its hot and cold junction ends and used in the measurement of elevated temperatures.

Tin   A chemical element having symbol Sn, formula weight 118.70, specific gravity 7.31, and melting point 231.85°C.

Tin sweat   Beads or exudations of a tin-rich low-melting phase found on the surface of or on risers of bronze castings, which are usually caused from absorption of hydrogen by the molten metal.

Tongs   Metal instrument with two legs joined by a hinger for grasping and holding things, e.g., crucible tongs.

Top board   A wood board on the cope half of the mold to permit squeezing the mold.

Transfer ladle   Container used to carry molten metal from the melting furnace to holding furnace, or from the furnace to pouring ladles.

Trim die   Die for shearing (or shaving) flash from a die-casting.

Trimming   Removing fins, gates, etc. from castings.

Trowel   Tool for sleeking, patching, and finishing a mold.

Tucking   Pressing sand with the fingers under the flask bars, around gaggers, and other places to insure firm placement.

Tumbling barrel   A revolving metal, wood box, or barrel in which castings are cleaned.

Tuyere   Opening through which the air blast enters the cupola.

Ultrasonic testing   Use of elastic waves of the same nature as sound, but of shorter wave length and higher frequency than those that affect the human ear (0.5–5 million cycles per sec), for detecting flaws in materials.

Undercut   Part of a mold or die requiring a drawback. (*See* Drawback.)

Vacuum degassing   Subjecting molten metal to a vacuum to remove deleterious gases such as hydrogen, oxygen, and nitrogen.

Vacuum melting   Melting, usually by induction heating, in a closed container which is subjected to a vacuum.

Vacuum refining   Vacuum melting to remove gaseous metal contaminants.

Veining   Surface defect on castings appearing as veins or wrinkles, which results from cracks in the sand due to elevated temperature conditions and occurs mostly in cores.

Vent   An opening in a mold or core to permit escape of steam and gases; it is also called a vent hole.

Vent rod   A piece of wire or bar to form the vents in sand.

Vent wax   Wax in rod shape placed in the core during manufacture. In the oven the wax is melted out, leaving a vent or passage.

Vibrator   A device which jars or vibrates the pattern (or matchplate) as it is withdrawn from the sand.

Viscosity   Resistance of a fluid substance to flowing.   A measurable characteristic for an individual substance at a given temperature and under definite conditions.

Wash   Casting defect resulting from erosion of sand by flowing metal. Also a term for coating materials applied to molds, cores, etc.

Wax   Class of substances of plant, animal, or mineral origin, insoluble in water, partly soluble in alcohol, ether, etc., and miscible in all proportions with oils and fats.   They consist of esters, free fatty acids, free alcohols, and higher hydrocarbons.   Common waxes are beeswax, bayberry, paraffin wax, ozokerite, ceresin, and carnauba.   Their mixtures are formed into rods and sheets and used for forming vents in cores and molds, repairing patterns, etc.

Weak sand   Sand lacking in the proper amount of bond.

Well (cupola)   Lower portion of a cupola, between the sand bottom and the slaghole, which forms a reservoir for the molten metal.

Wetting agent   Surface-active agent which by reducing surface tension of the wetting liquid causes a material to be wetted more easily.

Whirl gate   Gating system in which the metal enters a circular reservoir at a tangent, and so whirls around, leaving dirt and slag behind before passing into the mold cavity.

Whistlers   Small openings from isolated mold cavities to allow gases to escape easily. (*See* Vent.)

White cast iron   Cast iron in which substantially all the carbon is present in the form of iron carbide, and which has a white fracture.

Wood flour   Finely ground wood, usually hardwood, low in resin.

X-ray   Form of radiant energy with extremely short wave length which has the ability to penetrate materials that absorb or reflect ordinary light.

Yield   In production of castings, a value expressed as a percentage indicating the relationship of the weight of a casting to the total composite of the casting and its gating system.   For example, if the casting and gating system weigh 125 lb and the casting weighs 100 lb, the yield is 80%.

Zinc   A chemical element having symbol Zn, formula weight 65.38, specific gravity 7.140, and melting point 419.4°C.

Zircon   Natural zirconium silicate, $ZrSiO_4$, containing when pure 67.23% zirconium oxide, ZrO_4, and 32.77% silica, SiO_2, is used as a molding medium.

INDEX

INDEX